MS ACL — *The Association for* Mass Spectrometry: Applications to the Clinical Lab

MSACL 2019 EU

6th European Congress & Exhibits

Mass Spectrometry: Applications to the Clinical Lab

Salzburg, Austria

September 22-26, 2019

The Association is a non-membership, non-profit 501(c)(3) tax-exempt California Corporation with the mission of furthering education in the field of mass spectrometry.

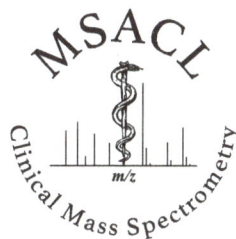

MSACL — Clinical Mass Spectrometry — m/z

clinical mass spectrometry
cms
an international journal

Get Connected with BadgerScan™
The Official Contact & Lead Capture App of MSACL

BadgerScan™

GET IT ON
Google play

Available on the iPhone
App Store

Restaurant Suggestions

There are a number of meals that you will be taking outside of the congress.
This is a great opportunity to see a little more of Salzburg and explore the city's culinary delights.

Walk	Name	Cost	Hours	Description
<5 m	The Heart of Joy	$$	8am-7pm, 8am-8:30pm Sundays	Vegetarian and vegan food.
	my Indigo	$	1am-10pm, 12pm-9pm Sundays	Asian Fusion, with fast take-away options and cosy interior for eating in. Top pick if you need something quick.
	Cafe Fingerlos	$$	7:30am-7:30pm	Best Breakfast in Town.
	IMLAUER Sky	$$$	9am-1:30am	Rooftop patio with a view. Bar for Drinks. Fancy food.
	Cafe Wernbacher	$$	9am-1:30am	
	Braurestaurant IMLAUER	$$	11am-12am	Austrian, traditional interior plus a beer garden, good for groups.
<10 m	Antichi Sapori	$$	11:30am-2pm, 5:30pm-9pm, Closed Wed,Thurs	Tasty Italian food and pizza.
	Gablerbrau	$$	11:30am-12am	Historic ambiance with traditional food.
	Cafe Sacher	$$$	7:30am-10:30pm	Nice for meal on the patio next to the river.
	Barenwirt	$$	11am - 11pm	Cozy and Traditional Austrian, fills up so arrive early, or make reservation.
	Zum fidelen Affen	$$$	12pm-2:30pm, 5pm-12am, Closed Sunday	Great for a nice evening meal on the patio. Make a reservation in advance. Small groups only.
<15 m	Die Wiesse	$$	10am-12am, Closed Sunday	Lovely Beer Hall and Garden, Traditional Austrian Food.
	Balkan Grill Walter	$	11am-7pm, 2pm-7pm Sundays	A favorite take-out place.
	Krimpelstatter	$$	11am-11pm. Hot food to 9pm. Closed Sun/Mon	Traditional Austrian food since 1584, Beer Garden, Good for groups up to 50 pax.
	Augustiner Braustubl	$	3pm-11pm	Historic Abby turned beer garden with some food stalls, good for very large groups. The place to be Thursday evening after the close of MSACL.
	Taj Mahal	$$	11:30am-1:45pm, 5:30-9:30pm	Good food in a lively atmosphere.
	Goldener Hirsch	$$$	Lunch & Dinner	Classic Salzburg.
	Fabi's Frozen Bio Yogurt	$	10am-9:30pm, 12pm-9pm Sundays	Frozen Yogurt in the Old Town! Yum
	Murphy's Law Pub	$	2pm -2am	After Dinner, you'll find folks from MSACL here for drinks and music.
<20 m	Stiftskeller St. Peter	$$$	11:30am-12am	Oldest in Europe (830 AD).
	M32	$$	9am-1am, Closed Monday	Take the elevator up at the Museum of Modernism. M32 offers breakfast, lunch, dinner or drinks in the city over the city.
	IceZeit	$	10am-10pm	Interesting ice cream flavors, some non-dairy, good quality.
<25 m	Steiglkeller	$$	11:30am-11pm	Great Beer Garden with a view over Salzburg. Historic building worth a look.
<35 m	Franziskischlössl	$$	11am-5pm, Closed Mon/Tues	There is a great outdoor cafe with drinks and a daily dish in the courtyard of the small castle at the top of this hill. Check opening times ahead. The walk is great!

Wednesday 16:30 in Mozart 1-5

15 years of Ambient Mass Spectrometry: From Amino Acid Clusters to Surgical Robotics
Zoltan Takats
Imperial College London

Plenary Lecture
Tuesday 14:15 in Mozart 1-5

Inflammatory Stories in Time and Space: Using Mass Spectrometry Imaging, Ion Mobility and High Throughput Lipidomics to Understand Human Disease
Jules Griffin
University of Cambridge, Cambridge, UK

Plenary Lecture
Thursday 17:15 in Mozart 4-5

Donor-derived Cell-free DNA as a Biomarker in Organ Transplantation
Michael Oellerich
Institute for Clinical Pharmacology, University Medicine Göttingen, Germany

Keynote Lecture : Session 1
Metabolomics
Wed 9:00 in Track 1 (Mozart 1-3)

From Spectrometric Data to Metabolic Networks: An Integrated Quantitative View of Cell Metabolism
Oscar Yanes
Rovira i Virgili University & IISPV

Keynote Lecture : Session 5
Glycomics
Thu 10:30 in Track 5 (Trakl Hall)

The Human Glycome Project - Exploring the New Frontier in Personalised Medicine
Gordan Lauc
University of Zagreb; Genos

Keynote Lecture : Session 2
Microbiology
Wed 11:00 in Track 6 (Doppler Hall)

Large-Scale Inference of Protein Tissue Origin in Sepsis Plasma Using Quantitative Targeted Proteomics
Johan Malmström
Lund University

Keynote Lecture : Session 6
Proteomics
Thu 13:15 in Track 2 (Mozart 4-5)

Antibody Sequencing and Quantitation
Theo Luider
Erasmus Medical Center

Keynote Lecture : Session 3
Proteomics
Wed 14:30 in Track 2 (Mozart 4-5)

The Development of Targeted Proteomic Assays, Attempting to Take Biomarkers from the Research Lab to the Clinic
Kevin Mills
University of London

Keynote Lecture : Session 7
Regulations
Thu 14:30 in Track 4 (Paracelsus Hall)

Impact of the New European IVD Regulation on Medical Laboratories - Opportunities and Challenges
Folker Spitzenberger
Technische Hochschule Lübeck

Keynote Lecture : Session 3
Breath Analysis
Wed 14:30 in Track 3 (Papageno Hall)

Selected Ion Flow Drift Tube MS SIFDT-MS for Real-Time Measurement of Trace Concentrations of Volatile Compounds in Breath and Culture Headspace
Patrik Španěl
J. Heyrovský Institute of Physical Chemistry

Keynote Lecture : Session 8
Tissue Imaging
Thu 15:45 in Track 2 (Mozart 4-5)

Imaging the Unimaginable with Imaging Mass Cytometry
Frits Koning
LUMC

Keynote Lecture : Session 4
Endocrinology
Thu 8:30 in Track 4 (Paracelsus Hall)

Rethinking Sex Steroids: Understanding the Clinical Relevance of 11-Oxygenated Androgens
Karl Storbeck
Stellenbosch University

Keynote Lecture : Session 8
Lipidomics
Thu 15:45 in Track 5 (Trakl Hall)

Skin Lipidomics in the Diagnosis and Treatment of Cutaneous Inflammation
Anna Nicolaou
Manchester University

Short Courses: Sun - Tue : September 22-24

Data Science 101
@ Trakl Hall (3rd Floor)
Breaking up with Excel: An Intro to the R Statistical Programming Language
Daniel Holmes, MD, William Slade, PhD, Stephen Master, MD, PhD

Data Science 201
@ Mozart 3 (Ground Floor)
Going Further With R: Tackling Clinical Laboratory Data Manipulation and Modeling
Patrick Mathias, MD, PhD, & Shannon Haymond, PhD

Glyco-Proteomics 101
@ Mozart 2 (Ground Floor)
Unravelling The Sweetness Of Life: Clinical Glyco(Proteo)Mics By Mass Spectrometry
Noortje de Haan, PhD & Guinevere Lageveen-Kammeijer, PhD

LC-MSMS 101
@ Paracelsus Hall (2nd Floor)
Getting Started with Quantitative LC-MS/MS in the Diagnostic Laboratory
Laura Owen, PhD, Michael Wright, Coral Munday & Grace van der Gugten

LC-MSMS 301
@ Mozart 4-5 (Ground Floor)
Development and Validation of Quantitative LC-MS/MS Assays for Use in Clinical Diagnostics
Brian Rappold & Chris Shuford, PhD

Lipidomics 101
@ Papageno Hall (Ground Floor)
Mass Spectrometry-based Lipidomics and Clinical Applications
Anne Bendt, PhD, Amaury Cazenave-Gassiot, PhD & Michael Chen, MD

Metabolomics 202
@ Doppler Hall (4th Floor)
Metabolomics: Approaches, Applications and Challenges
Julijana Ivanisevic, PhD & Elizabeth Want, PhD

Proteomic Microbiology 201
@ Trapp Zimmer Board Room (5th Floor)
Bottom-Up and Top-Down Proteomic Approaches for Bacterial Identification and Characterization, a Focus on MALDI-TOF and Advanced Technologies
Jean Armengaud, PhD & Stefan Zimmermann, MD

Scientific Committee

Elizabeth Want, PhD
Imperial College London
SciCom Chair

Ilaria Belluomo, PhD (Lead)
Imperial College London
Breath & VOC Analysis
Early Career Scientist Representative

Anne Bendt, PhD (Lead)
Singapore Lipidomics Incubator (SLING)
Lipidomics

Graeme Eisenhofer, PhD (Lead)
University of Dresden, Germany
Endocrinology

Flaminia Fanelli, PhD
University of Bologna, Italy
Endocrinology

Tom Fiers, MD
University of Ghent, Belgium
Endocrinology

Brian Keevil, PhD
University Hospital of South
Manchester, UK
Endocrinology

Laura Owen, PhD
University Hospital of South
Manchester, UK
Endocrinology & Practical Training

Manfred Wuhrer, Prof. dr. (Lead)
LUMC
Glycomics

Guinevere Lageveen-Kammeijer, PhD
LUMC
Glycomics
Early Career Scientist Representative

Gordan Lauc, PhD
University of Zagreb; Genos
Glycomics

Dirk Lefeber, PhD
Radboudumc
Glycomics

Oleg Mayboroda, PhD (Lead)
LUMC
Metabolomics

Julien Boccard, PhD
University of Geneva
Metabolomics
Early Career Scientist Representative

Warrick Dunn, PhD
University of Birmingham
Metabolomics

Julijana Ivanisevic, PhD
University of Lausanne
Metabolomics

Edward Moore, PhD (Lead)
University of Gothenberg, Sweden
Microbiology

Jean Armengaud, PhD
CEA-Marcoule, France
Microbiology

Simon Cameron, PhD
Queen's University Belfast
Microbiology
Early Career Scientist Representative

Stefan Zimmerman, MD
University Hospital Heidelberg, Germany
Microbiology

5

Scientific Committee, continued

Grace van der Gugten (Co-Lead)
Provincial Health Services Authority, Canada
Practical Training

Laura Owen, PhD (Co-Lead)
University Hospital of South Manchester
Practical Training & Endocrinology

Roland Geyer, PhD
Thermo Fisher Diagnostics
Practical Training

Judy Stone, PhD
UCSF
Practical Training

Michael Wright
LGC
Practical Training

Christa Cobbaert, PhD (Lead)
Leiden University Medical Center
Proteomics

Éva Hunyadi-Gulyás, PhD
Biological Research Centre
Proteomics

Lars Melholt Rasmussen, MD, DMSc
Odense University Hospital,
University of Southern Denmark
Proteomics

Renee Ruhaak, PhD
LUMC
Protemics
Early Career Scientist Representative

Jody van den Ouweland, PhD (Lead)
Canisius-Wilhelmina Hospital,
The Netherlands
Small Molecules / Tox / TDM

Ido Kema, Prof. Dr.
University Medical Center Groningen,
The Netherlands
Small Molecules / Tox / TDM

Marine Letertre, PhD
Imperial College London
Small Molecules / Tox / TDM
Early Career Scientist Representative

Michael Vogeser, Prof. Dr. med.
Institute of Laboratory Medicine,
Hospital of the University of Munich, Germany
Small Molecules / Tox / TDM

Tiffany Porta, PhD (Lead)
Maastricht MultiModal Molecular
Imaging (M4I) institute
Tissue Imaging

Isabelle Fournier, PhD
Université des Sciences et Technologies de Lille 1
Tissue Imaging

Ron Heeren, Prof. Dr.
Maastricht University
Tissue Imaging

Ólöf Gerdur Ísberg
University of Iceland / Imperial College London
Tissue Imaging
Early Career Scientist Representative

Kristina Schwamborn, MD, PhD
Institute of Pathology, TU Munich
Tissue Imaging

Raf Van de Plas, PhD
Delft University of Technology
Tissue Imaging

Nicola Gray, PhD
University of Reading
Early Career Scientist Committee Chair

Tuesday

615
715

Early-Career Scientists Committee (ECSC) – Morning City Run
@ *Entrance Foyer*
Chair: Nicola Gray
Everyone is welcome to join the ECSC for a morning 5 km run in Salzburg. An ideal way to see some of the city while working off the Austrian cuisine! All abilities welcome.

800
1600

Place Posters
@ *1st Floor Exhibit Hall*
ALL posters should be placed by 16:00.

800
900

Welcome Coffee
@ *Entrance Foyer*
Enjoy coffee, a pastry and a chat with colleagues before the day starts.

900
1300

Get the Basics
@ *Papageno Hall*
Chairs: Roland Geyer & Anne Schmedes

This series of Primers on emerging fields within Clinical Mass Spec is free to attend for ALL registrants of the MSACL Congress or Short Courses, including Partial Pass holders.

Take aways for each session include:
1. Be able to define what the field is, why it matters, and why you personally should care. What is the clinical relevance?
2. Understand what role mass spectrometry plays in this field. Where does mass spec fit in to the big picture of the field?
3. Define any terminology that is specific to this field.
4. Describe how it works, what are the methods and workflows used when studying this field and/or how is it implemented in clinical labs.
5. Identify any challenges to implementation/adoption, where do the opportunities lie?

09:00 - 09:40 **Glycomics** with Manfred Wuhrer
09:50 - 10:30 **Mass Spec Imaging** with Tiffany Porta
10:40 - 11:20 **Lipidomics** with Anne Bendt
11:30 - 12:10 **Data Science** with Shannon Haymond
12:20 - 13:00 **Breath Analysis** with Ilaria Belluomo

13:10
13:50

Discussion Group: Study Design and Logistics for Metabolomics Studies
@ *Papageno Hall*
Chair: Jerzy Adamski

Study design is a very critical step in metabolomic experiment. It has to ensure that phenotype resolution, metabolite coverage, quality-driving resources and logistic processes are considered. During the session we will interactively discuss steps like type of the study, consideration of confounders, power calculation, pre-analytical requirements, sample storage, randomization, metabolite analytics to be chosen and replication considerations.

1400
1415

Welcome, Introduction & Orientation
@ *Mozart 1-5*
Chair: Chris Herold

Opening Plenary Lecture
@ *Mozart 1-5*
Chair: Anne Bendt

1415
1500

Inflammatory Stories in Time and Space: Using Mass Spectrometry Imaging, Ion Mobility and High Throughput Lipidomics to Understand Human Disease
Jules Griffin
Computational and Systems Medicine, Surgery and Cancer, Imperial College London, London and Department of Biochemistry, University of Cambridge, Cambridge, UK

A central aspect of the development of many of the pathologies associated with the metabolic syndrome is a chronic progression of inflammation in the affected tissues. This is in part driven by lipid remodelling in the cell membrane and the production of pro-inflammatory lipid mediators produced from polyunsaturated fatty acids such as arachidonic, docosahexaenoic and eicosapentaenoic acids. To explore lipid remodelling during the development of non-alcoholic fatty liver disease we have applied MALDI-based mass spectrometry imaging (MSI) to examine both human tissue and animal models of the disease progression. Using a combination of high fat feeding and genetic modification (the ob/ob mouse which lacks leptin) to cause hepatic steatosis with and without inflammation, MSI shows that one of the events associated with disease progression is a lipid remodelling of phosphatidylcholines (PCs), and in particular, a reduction in arachidonic acid containing PCs. We have also developed a ultra-high performance liquid chromatography ion mobility mass spectrometry-based method to profile known and novel lipid mediators, using a KNIME workflow to process the data and annotate the detected lipids, in part relying on collision cross-section values for these species to aid assignments. This will be illustrated in following the time course of lipid changes in thrombin activated human platelets.

TUESDAY

1515 1615	**State of the Science Address** *@ Mozart 1-5*

Chairs: Elizabeth Want and Roland Geyer
Speakers: Ilaria Belluomo, Graeme Eisenhofer, Manfred Wuhrer, Anne Bendt, Oleg Mayboroda, Edward Moore, Lars Melholt Rasmussen, Jody van den Ouweland, Tiffany Porta, Grace van der Gugten
Presented by MSACL EU Scientific Committee, this address will provide an overview of the applications and technologies currently being used in Clinical Labs, and a clear view of the development pipeline. It will highlight applications expected to be available in the near-future, as well as emerging applications, and key contributors. Exemplary talks, posters, and people present at the congress will be identified, enabling you to optimize your learning path and more effectively target potential network connections. Whether you are new to Clinical Mass Spectrometry, or a seasoned veteran, the State of the Science address should be on your agenda.

1615 1630	**Intermission** *@ Entrance Foyer*

1630 1700	**Poster Lightning Talks** *@ Mozart 1-5* *Chairs:* Ilaria Belluomo and Julien Boccard *Judges:* Simon Cameron, Eva Hunyadi-Gulyas, Guinevere Lageveen-Kammeijer, Renee Ruhaak, Will Slade, Karl Storbeck, Grace van der Gugten, Elizabeth Want Poster Contest Semi-Finalists will be selected to present their pitch from the plenary podium within 90 seconds and with 1 slide (PDF format).

1700 1715	**Exhibitor Lightning Talks** *@ Mozart 1-5* *Chair:* Nicola Gray *Presenting Companies:* Bioinformatics Solutions, Cambridge Isotope Labs, Chromsystems, Merck, Shimadzu, Tecan, Thermo Fisher Scientific, UTAK Laboratories, Waters Learn what's new from select vendors.

1715	**EXHIBITS OPEN** *@ 1st Floor Exhibit Hall*

1715 1930	**Exhibitor Reception** *@ 1st Floor Exhibit Hall* Dinner served.

1800 1900	**Meet-a-Mentor : Booth Tours** *@ 1st Floor Exhibit Hall : Meet-a-Mentor Rally Point* *Chair:* Renee Ruhaak *Mentors:* David Herold, Laura Owen, Grace van der Gugten, Michael Wright Join an expert mentor as they tour the booths and learn more about technologies available, what experts are looking for during the exchange, and how they interact with the booth vendors. Increase your network and your knowledge of products and services available.

1930 2030	**FeMS: Networking Reception** *@ Mozart 1-3* *Chairs:* Anne Bendt, Grace van der Gugten, Margrét Þorsteinsdóttir Please join us for this inaugural FeMS event! Over wine and cheese, a few thought-provoking questions will be posed. Afterwards we'll have a discussion group at 20:30 (see below). Organized by Females in Mass Spectrometry (FeMS), an initiative to bring together women in the mass spectrometry field.	**Early-Career Scientists (ECS) - Career Options in Clinical Mass Spectrometry Discussion** *@Mozart 4-5* *Chair:* Nicola Gray This discussion group is aimed at early-career scientists and those involved with training and recruitment of the next generation. The goal is to create awareness on career development and planning. The panel includes speakers from academia, industry and entrepreneurship and will explore transitioning between sectors. Speakers include **Christiane Auray-Blais** (Univ of Sherbrooke, Canada), **Brian Keevil** (Univ of Manchester, UK), **Flaminia Flanelli** (Univ of Bologna, Italy), **Christian Scherling** (Tecan, Germany), and **Will Slade** (LabCorp, USA) who will discuss their own professional career paths.

2030 2130	**FeMS : Discussion Group : How to Gain Visibly in the Field of Mass Spectrometry** *Chair:* Rita Horvath *with panelists:* Gwen McMillin, Margrét Þorsteinsdóttir, Julijana Ivanisevic, I-Lin Tsai & Elizabeth Want Thought-provoking questions, followed by a discussion with women at various stages in their careers, including an open period for audience questions. Exploring themes such as: (i) Best practices to achieve more visibility; (ii) How can the mass spec community participate in supporting its male and female members? (iii) How can we encourage each other to succeed? What questions do you have? Open to all! Additionally, the "Rowena Shaw Avanti Award for Women in Mass Spectrometry", a $2000 USD discretionary award funded by Avanti Polar Lipids, will be introduced.

Wednesday

800 845	**Corporate Workshop(s)**

Merck
@ Mozart 4-5 (Track 2)
Enhancing Accuracy: Matrix Selection, Internal Standard Selection, and Sample Clean-Up Solutions for Clinical Mass Spectrometry
Geoffrey Rule,PhD, Merck

Thermo Fisher Scientific
@ Papageno (Track 3)
1. Achieve Superior Confidence in Your LC-MS Analysis, *Claudio DeNardi, Thermo Fisher*

2. Trying to stay one-step ahead – how High-Resolution Accurate Mass LCMS has enabled early detection of fentanyl analogues (including carfentanil) and synthetic cannabinoids in the UK, *Simon Hudson, LGC, UK*

SCIEX
@ Paracelsus (Track 4)
Advanced mass spectrometry techniques provide deeper access into the vitamin D metabolome
Pascal Schorr, Humboldt-University, Berlin

Scientific Session 1

	Track 1 Mozart 1-3 Metabolomics Keynote Chair: Oleg Mayboroda 2nd Chair: H. Koguna	Track 2 Mozart 4-5 Proteomics: Reinventing Old Biomarkers Chair: TBA	Track 3 Papageno Hall Quality Control Chair: Rita Horvath	Track 4 Paracelsus Hall Small Molecules: Metabolomics in Toxicology Chair: Jody van den Ouweland 2nd Chair: Juliane Fagotti	Track 5 Trakl Hall Lipidomics: Applications for Doping, Nutrition and Lab Med Chair: Erika Dorado 2nd Chair: Panagiotis Vorkas	Track 6 Doppler Hall Breath Analysis and VOC in Cancer Chair: Kseniya Dryahina
900 920	◎ From Spectrometric Data to Metabolic Networks: An Integrated Quantitative View of Cell Metabolism Oscar Yanes *CIBER & URV & IISPV*	◎ HILIC-LC-MRM-MS Enables the Quantitation of PSA and its Major Glycoforms to Improve Prostate Cancer Diagnostics L.Renee Ruhaak *Leiden University Medical Center*	◎ A Proposal to Standardize the Description of LC-MS -based Measurement Methods in Laboratory Medicine Michael Vogeser *Hospital of the University of Munich (LMU)*	◎ UPLC-MS/MS-based Assays for Diagnosis and Therapeutic Monitoring in Patients with APRT Deficiency Unnur Arna Thorsteinsdottir *University of Iceland* Young Investigator Grant	◎ Development and Validation of a Gas-Chromatography Mass Spectrometry Method for the Analysis of Testosterone Esters in Biological Matrices Michele Iannone *Laboratorio Antidoping FMSI, Rome, Italy* Young Investigator Grant	◎ Investigating the Relationship between TRIM44 and P53 Metabolic Pathways in Human Colon Cancer Cells in a 3D Model, Using PTR-ToF-MS Katerina-Vanessa Savva *PhD, Imperial College London* Young Investigator Grant
920 940	... Keynote ...	◎ Proteomic Analysis of Prostate Cancer Biopsies Gábor Tóth *MS Proteomics Research Group, RCNS-HAS* Young Investigator Grant	◎ Metrological Traceability of Lp(a) Requires Defining the Measurand and Introducing Molar Units Christa Cobbaert *Leiden University Medical Center*	◎ Investigation of Bacterial β-Glucuronidase Inhibition on the Metabolic Profile of Mice Using a Combined Platform of NMR and LCMS-based Untargeted Metabolomics Marine Letertre *Imperial College London* Young Investigator Grant	◎ Determination of Serum Triglyceride by Isotope Dilution LC-MS/MS Weiyan Zhou *NCCL，Beijing Hospital*	◎ VOC-based Real-Time Breast Cancer Diagnostic Using LTP-MS Flore Herve *PhD student, PRISM U1192* Young Investigator Grant
940 1000	... Keynote ...	◎ The Analysis of Alpha-1-Antitrypsin Glycosylation in HCC Patient Serum Haidi Yin *The Hong Kong Polytechnic University*	◎ Rigorous Quality Management Enables Long-Term Robustness of MS-based Protein Quantitation Nico Smit *Leiden University Medical Center*	◎ Development of a Liquid Chromatography Quadrupole Time-of-Flight Mass Spectrometry (LC-QTof MS) Method for the Screening of Antihypertensive Drugs in Urine Alexander Lawson *University Hospitals Birmingham NHS FT*	◎ Whole Blood Omega-3 Index by Gas Chromatography Dependence on Age and Gender in Caucasian Population Vasiliy Yurasov *Scientific laboratory "Chromolab"*	◎ Volatile Organic Compound (VOC) Profiling in Liver Disease via Selected-Ion Flow-Tube Mass Spectrometry (SIFT-MS) Michael Hewitt *Imperial College London* Young Investigator Grant

◎ = Emerging. More than 5 years before clinical availability. (35.32%)
◎ = Expected to be clinically available in 1 to 4 years. (32.77%)
◎ = Clinically available now. (31.91%)

1000
1100

Poster Session 1
@ 1st Floor Exhibit Hall
Selected posters to be attended for 1 hour.

Scientific Session 2

	Mozart 1-3	Mozart 4-5	Papageno Hall	Paracelsus Hall	Trakl Hall	Doppler Hall
	Metabolomics: Clinical Apps I *Chair: Oleg Mayboroda* *2nd Chair: Chung-Ho Lau*	**Proteomics: Metrological Traceability and Absolute Quantification** *Chair: Christa Cobbaert*	**Practical Training: Traceable Clinical Measurement** *Chair: Roland Geyer*	**Endocrinology: Serum Steroids** *Chair: Karl Storbeck* *2nd Chair: Elizabeth Baranowski*	**Clinical Glycomics I** *Chair: Guinevere Lageveen-Kammeijer* *2nd Chair: Gábor Tóth*	**Microbiology Keynote** *Chair: Roger Karlsson* *2nd Chair: Petra Paizs*
1100 **1120**	◎ Metabolomics Analysis of Blood Plasma for Early Diagnosis of Uterus Leiomyoma Recurrences Vladimir Frankevich *Research Center for Obstetrics and Gynecology*	◎ Development of an Immunoprecipitation Method for the Analysis of Intact Parathyroid Hormone (PTH) and Related Forms by LC-MS/MS Jordi Farré-Segura *University of Liège* Young Investigator Grant	◎ Traceable Clinical Measurement Chris Hopley *NML*	◎ The Harmoster Initiative: Preliminary Results on the Comparability of Circulating Steroid Measurement Among Ten European Laboratories Using LC-MS/MS Flaminia Fanelli *University of Bologna*	◎ Comprehensive Glycopeptide Profiling in Blood Plasma for Clinical Applications Melissa Baerenfaenger *Radboudumc*	◎ Large-Scale Inference of Protein Tissue Origin in Sepsis Plasma Using Quantitative Targeted Proteomics Johan Malmström *Department of Clinical Sciences*
1120 **1140**	◎ Metabolomics Applied to Coronary Clinical Burden: Use of Mass Spectrometry to Re-Stratify Asymptomatic Subjects of Intermediate Cardiovascular Risk Aline Martins *Universidad CEU San Pablo Madrid*	◎ Multiplexing Kidney Injury and Kidney Specific Biomarkers in Urine Using MS-based Bottom-Up Proteomics: A Robust and Feasible Approach Tirsa van Duijl *Leiden University Medical Center (LUMC)* Young Investigator Grant	... Extended Session ...	◎ Multidimensional Diagnostics with Machine Learning for Primary Aldosteronism Graeme Eisenhofer *Technische Universität Dresden*	◎ Ultrahigh Resolution MS Profiling of Plasma N-Glycans in Large Type 2 Diabetes Cohorts Elham Memarian *Leiden University Medical Center* Young Investigator Grant	... Keynote ...
1140 **1200**	◎ Metabotyping Burn Injury Using UPLC-MS Coupled with Microdialysis Elizabeth Want *Imperial College London*	◎ The Combined Use of Lys-C and Trypsin Provides Better Digestion Efficiency for MS-based Protein Quantitation Fred Romijn *Leiden University Medical Center*	... Extended Session ...	◎ An Ultrasensitive High-Throughput LCMS/MS-Method for Estradiol and Estrone in the Sub-Picomolar Range Bjørn-Erik Bertelsen *Hormone Laboratory, Haukeland University Hospital*	◎ IgG and IgA Glycopeptide Characterization by LC-MS Reveals Associations with Inflammatory Bowel Disease Subtypes and Behavior Noortje de Haan *LUMC*	... Keynote ...

1200
1330

Lunch
@ 1st Floor Exhibit Hall

1230
1315

Troubleshooting Poster Rounds : Part II
@ 1st Floor Exhibit Hall with Chair, Grace van der Gugten

12:30 at poster #24a
The Hurdles of Developing an LC-MS/MS Assay for Desmosine, a Biomarker for Elastin Degradation
Jody van den Ouweland
Canisius-Wilhelmina Hospital

12:45 at poster #24b
Fast and Efficient Method for Detection and Quantification of Catecholamines in Human Plasma by LC-MS/MS - Troubleshooting
Chiara Calaprice
University Hospital of Liege

13:00 at poster #24c
Matrix Dilemmas in LC-MS/MS: The Use of Appropriate Calibrators and Internal Standards to Facilitate Acceptable Criteria for Quantification
Tania Grobler
North-West University

WEDNESDAY

1330–1415 Corporate Workshop(s)

Shimadzu
@ Mozart 4-5 (Track 2)
1. MALDI-MS technology combined with automated interpretation software, a solution adapted to high throughput screening: Sickle cell disease screening in newborns as a first application
2. Robustness and reliability of the novel pre-analytical filtration system CLAM-2030 and LCMS-8050
3. Rethinking the capability and workflows in clinical metabolomics and biomarker research using the LCMS-9030 Q-TOF technologies

Thermo Fisher Scientific
@ Papageno (Track 3)
Laboratory-Developed Mass Spectrometry Testing vs Regulatory-Approved Tests: Pros and Cons
Part 1: Balancing the pros and cons of Mass Spec-based Lab-Developed-Tests as compared to Regulatory-Approved Tests under the new IVD Regulation 2017/746 in Europe
Part 2: Cascadion™ SM 25-Hydroxy Vitamin D and Immunosuppressants Panel—Assays on the Cascadion™ SM Clinical Analyzer

Waters
@ Paracelsus (Track 4)
Mass Spectrometry advantages for Therapeutic Drug Monitoring: a closer look at teicoplanin
Brian Keevil, Professor at University Hospital Southern Manchester, UK

Agilent Technologies
@ Trakl (Track 5)
Fast Track your Cannabinoid Urine Analysis using MassHunter StreamSelect LC-MS System
Moritz Wagner, Product Specialist, Agilent Technologies

Scientific Session 3

	Mozart 1-3	Mozart 4-5	Papageno Hall	Paracelsus Hall	Trakl Hall	Doppler Hall
	Metabolomics: Clinical Apps II *Chair: Nicola Gray*	**Proteomics Keynote** *Chair: Christa Cobbaert*	**Breath Analysis Keynote: Patrik Španěl** *Chair: Ilaria Belluomo*	**Small Molecules: From Vitamins, Drugs to Chemical Warfare** *Chair: Michael Vogeser*	**Lipidomics in Cancer** *Chair: Vladimir Frankevich 2nd Chair: Kenji Nakayama*	**MALDI for the Clinical Laboratory** *Chair: David Herold*
1430 1450	Assessing the Metabolome of Preterm Newborns: Findings from a Danish Population-based Study Madeleine Ernst *Statens Serum Institut*	The Development of Targeted Proteomic Assays, Attempting to Take Biomarkers from the Research Lab to the Clinic Kevin Mills *University College London*	Selected Ion Flow Drift Tube Mass Spectrometry SIFDT-MS for Real-Time Measurement of Trace Concentrations of Volatile Compounds in Breath and Culture Headspace Patrik Španěl *J. Heyrovský Institute of Physical Chemistry*	Development of an LC-MS/MS Assay for Biotin Quantitation to Support Research on Biotin Interference of Immunoassays Antoinette IJpelaar *CWZ Hospital* Young Investigator Grant	MALDI-TOF/MS Analyses of Urinary Phospholipids and Lysophospholipids as Potential Diagnostic Biomarkers for Prostate Cancer Xin Li *Kyoto University* Young Investigator Grant	Optimization of MALDI-based Techniques for the Rapid Diagnostics of Sepsis Miriam Cordovana *A.O.U. Policlinico Sant'Orsola-Malpighi*
1450 1510	Podocyturia Evaluation in Women with Preeclampsia and Fabry Disease Patients Using a Tandem Mass Spectrometry Approach Christiane Auray-Blais *Université de Sherbrooke*	... Keynote ...	... Keynote ...	LC-MS/MS for Peptide-Adduct Analysis to Verify Poisoning with Chemical Warfare Agents and Pesticides Harald John *Bundeswehr Institute of Pharmacology & Toxicology*	Can Rapid Evaporative Ionization Mass Spectrometry (REIMS), the Intelligent Surgical Knife (iKnife) Provide a Point-of-Care Diagnosis for Endometrial Cancer Diana Marcus *Imperial College London* Young Investigator Grant	MALDI Chip Technology for in situ Detection of Human Procalcitonin Petr Pompach *Institute of Microbiology*
1510 1530	Exploratory Metabolomics of Urine and Plasma to Identify Novel Pharmacodynamic Biomarkers in a Phase I Clinical Trial of AZD3965 Alexandros Siskos *Imperial College London* Young Investigator Grant	... Keynote ...	... Keynote ...	Analysis of Fentanyl and Metabolite in Clinical and Post-Mortem Samples Karen Scott *Arcadia University*	Lipidomic Study of Cell Lines Reveals Differences between Breast Cancer Subtypes Finnur Freyr Eiriksson *University of Iceland*	Potential of TiO2 Nanoparticles for Mass Spectrometric Detection and Quantification of Biologically-Relevant Small Molecules Marijana Petkovic *University of Madeira, CQM-Centre for Chemistry*

1530
1630
Poster Session 2
@ *1st Floor Exhibit Hall*

1530
1630
Meet-a-Mentor : Poster Tours
@ *1st Floor Exhibit Hall : Meet-a-Mentor Rally Point*
Chair: Renee Ruhaak
Mentors: Isabelle Fournier, David Herold, Anna Nicolaou, Brian Rappold

Distinguished Contribution Award Lecture
@ *Mozart 1-5*
Chair: Elizabeth Want

1630
1730

15 years of Ambient Mass Spectrometry: From Amino Acid Clusters to Surgical Robotics
Zoltan Takats
Imperial College London

Professor Takats obtained his PhD from Eötvös Loránd University, Budapest, Hungary. He worked as a post-doctoral research associate at Purdue University, Indiana, USA. After returning to Hungary, he served as Director of the Cell Screen Research Centre and also as Head of Newborn Screening and Metabolic Diagnostic Laboratory at Semmelweis University, Budapest.

Professor Takats was awarded a Starting Grant by the European Research Council in 2008 and he subsequently became a Junior Research Group Leader at Justus Liebig University, Gießen, Germany. He moved to the United Kingdom in 2012 and is currently a Professor of Analytical Chemistry and Director of Mass Spectrometry Research within Division of Computational Systems and Medicine at Imperial College London.

Professor Takats has pursued pioneering research in mass spectrometry and he is one of the founders of the field of 'Ambient Mass Spectrometry'. He is the primary inventor of six mass spectrometric ionisation techniques, including Desorption Electrospray Ionization and Rapid Evaporative Ionization Mass Spectrometry.

He was the recipient of the prestigious Mattauch-Herzog Award of the German Mass Spectrometry Society and the Hungarian Star Award for Outstanding Innovators. He is the founder of Prosolia Inc, Medimass Ltd and Massprom Ltd, all companies pursuing analytical and medical device development. Professor Takats has published over 100 peer-reviewed articles in the fields of analytical chemistry and translational medicine.

1730
1900
Exhibitor Reception
@ *1st Floor Exhibit Hall*
Wine and Light Appetizers served in Exhibit Hall.

1800
1900
Meet-a-Mentor : Office Hours Forum
@ *1st Floor Cafe : Meet-a-Mentor Rally Point*
Chair: Renee Ruhaak
Mentors: Christiane Auray-Blais, Anne Bendt, Shannon Haymond, David Herold, Daniel Holmes, Patrick Mathias, Anna Nicolaou, Brian Rappold, Chris Shuford, Will Slade, Grace van der Gugten, Elizabeth Want
Have a question from the congress that you have been itching to ask? Or a problem from work that you want to get feedback on? Need career advice? Sign up at the registration desk for 15-min blocks to share time and ideas with domain experts.

1900
Enjoy the City
@ *Your Choice*
Plan to join a few colleagues for dinner in Salzburg tonight. Some places to try:
1. Braurestaurant IMLAUER (Austrian, traditional interior plus a beer garden, good for groups, close to congress)
2. Krimpelstatter (Traditional Austrian food since 1584, Beer Garden, serving hot food to 9pm, open to 11pm)
3. Steiglkeller (Historic building with a great view over the Old Town, near the Fortress)
4. Gablerbrau (Historic ambiance with traditional food)
5. Cafe Wernbacher (Austrian, great for a rainy day with original 1950s decor, open 9am-midnight daily, near congress)
6. Die Weisse
7. Indian Restaurant Taj Mahal

For a special meal (reservation advised):
1. Restaurant Goldener Hirsch
2. Stiftskeller St. Peter
3. Zum fidelen Affen
4. M32

After dinner:
1. Murphys Law Irish Pub (Common place to meet MSACL folks for drinks and music).

Thursday

800 830	**Welcome Coffee** @ *Entrance Foyer* Enjoy coffee, a pastry and a chat with colleagues before the day starts.

Scientific Session 4

	Track 1 Mozart 1-3	Track 2 Mozart 4-5	Track 3 Papageno Hall	Track 4 Paracelsus Hall	Track 5 Trakl Hall	Track 6 Doppler Hall
	Metabolomics: Clinical Applications III *Chair: Christiane Auray-Blais* *2nd Chair: Vitaliy Chagovets*	**Proteomics: High Throughput Methodology** *Chair: Éva Hunyadi-Gulyás*	**Practical Training** *Chair: Michael Wright*	**Endocrinology Keynote** *Chair: Brian Keevil*	**Clinical Glycomics II** *Chair: Valeriia Kuzyk*	**Data Science** *Chair: Stephen Master*
830 850	The Use of Rapid Evaporative Ionisation Mass Spectrometry (REIMS) as a Real-Time Bedside Test in Cervical Cancer Screening Maria Paraskevaidi *Imperial College London* Young Investigator Grant	Maximize the Output of Routine Proteome Analyses by Using Micro Pillar Array Column Technology Christof Mitterer *PharmaFluidics NV*	Multi-Analyte Methods: A Free Lunch or a Frightening Bill? Coral Munday *LGC Ltd*	Rethinking Sex Steroids: Understanding the Clinical Relevance of 11-Oxygenated Androgens Karl Storbeck *Stellenbosch University*	Successful Faecal Microbiota Transplantation for Recurrent Clostridioides difficile Infection Associates with Decreased Complexity of the Serum N-Glycome Maja Pučić Baković *Genos Glycoscience Research Laboratory*	A Chromatogram Says a Thousand Numbers: Data, Decisions, and Directions William Slade *Laboratory Corporation of America Holdings*
850 910	Clinically Relevant Metabolites of the Human Gut Microbiota Zdenek Spacil *Masaryk University*	DirectMS1: A High-Throughput Analysis for Identification and Quantification of 1000 Proteins in 5 Minutes Mark Ivanov *INEPCP FRCCP RAS* Young Investigator Grant	... Extended Session ...	... Keynote ...	Novel Glycopeptide Analysis Strategy for Prostate-Specific Antigen in Seminal Plasma of Infertile Men Wei Wang *Leiden University Medical Center* Young Investigator Grant	Chemometric Strategies for Sensitive Regions of Interest Annotation in Complex Imaging Mass Spectrometry Data Patrick Wehrli *University of Gothenburg*
910 930	Exploring the Interactions between Dietary Polyphenols, the Gut Microbiome and Human Health Nicola Gray *University of Reading, UK* Young Investigator Grant	Comparison of MRM- and DIA-based Protein Quantification Using a 500 Protein Blood Panel in Plasma Samples Demonstrates Linearity between Methods Claudia Escher *Biognosys AG*	... Extended Session ...	... Keynote ...	Colorectal Cancer Cell Lines Reveal Striking Isomeric Diversity of Human Cancer O-Glycome Katarina Madunic *LUMC* Young Investigator Grant	TBA

930 1030	**Poster Session 3** @ *1st Floor Exhibit Hall* Selected posters to be attended for 1 hour. Refer to program for posters attended during this period.

930 1030	**Poster Contest Finalist Interviews** @ *1st Floor Exhibit Hall* *Chairs:* Ilaria Belluomo and Julien Boccard *Judges:* Simon Cameron, Eva Hunyadi-Gulyas, Guinevere Lageveen-Kammeijer, Renee Ruhaak, Will Slade, Karl Storbeck, Grace van der Gugten, Elizabeth Want

Poster Contest Finalists to attend posters and meet with poster Judges. Judges may arrive in groups or individually.

THURSDAY

1030 1130	**Exhibitor Feedback Discussion** @ *1st Floor Cafe* All exhibitors are welcome to attend this discussion with the organizers and provide feedback.

Scientific Session 5

	Mozart 1-3	Mozart 4-5	Papageno Hall	Paracelsus Hall	Trakl Hall	Doppler Hall
	Metabolomics: Data Analysis Workflows *Chair: Julien Boccard* *2nd Chair: Víctor González Ruiz*	**Proteomics: Pre- analytical Considerations for Patient and Specimen Preparation** *Chair: Lars Rasmussen*	**Practical Training: Assessing the Quality and Robustness of LC-MS/MS assays** *Chair: Laura Owen*	**Small Molecules: From Basic to Novel Approaches in TDM** *Chair: Marine Letertre*	**Glycomics Keynote** *Chair: Manfred Wuhrer*	**Microbiology: Tandem MS-MS Proteomics and Clinical Diagnostics** *Chair: Jean Armengaud* *2nd Chair: Miriam Cordovana*
1030 1050	Plasma Metabolomics Profiles Associated with Endothelial Health and Dysfunction and their Influence on Endothelial Metabolism Óttar Rolfsson *University of Iceland*	Precise Quantitative Serum LC-MS/MS Profiling: The Impact of Sample Preparation and Sample Source on Biomarker Discovery Studies Oleksandr Boychenko *Thermo Fisher Scientific*	Assessing the Quality and Robustness of LC-MS/MS Assays James Hawley *Manchester University NHS Foundation Trust*	Application of a Molecular Networking Approach for Therapeutic Drug Monitoring (TDM) and Toxicology Emmanuel Bourgogne *Université de Paris, Faculté de Pharmacie, UMR8038*	The Human Glycome Project - Exploring the New Frontier in Personalised Medicine Gordan Lauc *University of Zagreb*	Discovery of Species-Unique Peptides as Candidate Peptide Biomarkers for Respiratory Tract Pathogens Using Tandem Mass Spectrometry-based Proteotyping Roger Karlsson *Sahlgrenska University Hospital*
1050 1110	KIMBLE: A Versatile Visual NMR Metabolomics Workbench in KNIME Aswin Verhoeven *Leiden University Medical Center*	Proteomic Discovery and Validation of the Confounding Effect of Heparin Administration on the Analysis of Candidate Cardiovascular Biomarkers Hans Christian Beck *Odense universitetshospital*	... Extended Session ...	Potential of MALDI MS for Detection of [Ru(η5-C5H5) (PPh3)2Cl] Metabolites in Patient Body Fluids – a Preliminary Study Nádia Nunes *CQM, University of Madeira* Young Investigator Grant	... Keynote ...	Quick Pathogen Identification and Antibiogram for the Clinics by MS/MS Proteotyping Jean Armengaud *CEA*
1110 1130	Computational Analysis of Mass Spectrometry Data for Standardised Diagnosis of Inborn Disorders of Steroidogenesis Elizabeth Baranowski *IMSR, University of Birmingham* Young Investigator Grant	The Predictive Potential of Salivary Proteomics for Characterization of the Phases of Allogeneic Hematopoietic Stem Cell Transplants Milena Souza *International Research Center, A.C.Camargo Cancer* Young Investigator Grant	... Extended Session ...	Therapeutic Drug Monitoring of Clofazimine in Indian MDR-TB Patients Prerna Chawla *P.D. HINDUJA HOSPITAL & MRC* Young Investigator Grant	... Keynote ...	Robust Accurate Identification and Biomass Estimates of Microorganisms via Tandem Mass Spectrometry Yu Yi-Kuo *National Center for Biotechnology Information*

1130 1315	**Lunch** @ *1st Floor Exhibit Hall*

THURSDAY

1200 1300	**Poster Session 4** @ *1st Floor Exhibit Hall*

1200 1300	**Meet-a-Mentor : Poster Tours** @ *1st Floor Exhibit Hall : Meet-a-Mentor Rally Point* *Chair:* Renee Ruhaak *Mentors:* Isabelle Fournier, David Herold, Brian Rappold

1300	**Remove Posters** @ *1st Floor Exhibit Hall*

1315	**EXHIBITS CLOSED** @ *1st Floor Exhibit Hall*

Scientific Session 6

	Mozart 1-3	Mozart 4-5	Papageno Hall	Paracelsus Hall	Trakl Hall	Doppler Hall
	Metabolomics of Model Systems *Chair: Petra Paizs*	**Proteomics Keynote** *Chair: Christa Cobbaert*	**Practical Training: Calibration and Metrology for Protein MS tests** *Chair: Grace van der Gugten*	**Endocrinology: Miscellaneous** *Chair: Flaminia Fanelli*	**MS Imaging: Method Development** *Chair: Isabelle Fournier*	**Breath Analysis and VOC: New Technology** *Chair: Patrik Spanel*
1315 1335	⊙ LC- and CE-MS-based Workflow for Metabolic Read-Across of New Toxicants in Neuroinflammation Víctor González-Ruiz *University of Geneva* Young Investigator Grant	⊙ Antibody Sequencing and Quantitation Theo Marten Luider *Erasmus MC*	⊙ Calibration and Metrology for Protein MS Tests L.Renee Ruhaak *Leiden University Medical Center*	⊙ Validation of a Rapid LC-MS/MS Method with Simple Solid Phase Extraction for Analysis of Catecholamines and their Metabolites in Plasma Songlin Yu *Peking Union Medical College Hospital*	⊙ Imaging Spatial Abeta Plaque Aggregation Dynamics in Evolving AD Pathology Using iSILK Wojciech Michno *Sahlgrenka Academy at the University of Gothenburg* Young Investigator Grant	⊙ Real-Time Therapeutic Monitoring of Valproic Acid in Exhaled Breath Kapil Dev Singh *University of Basel*
1335 1355	⊙ Metabolomic Analysis of Human Atherosclerotic Plaques Reveals a Pathway of Foam Cell Apoptosis in Advanced Atherosclerosis Panagiotis Vorkas *Imperial College London* Young Investigator Grant	... Keynote ...	... Extended Session ...	⊙ A Novel LC-MS/MS Method for the Quantification of Five Androgens in Saliva Malcolm McTaggart *Manchester University NHS Foundation Trust*	⊙ High-Throughput Automated Tissue Imaging by DESI-MS, Three-Dimensional Tissue Imaging or Large Sample Cohort Mapping Emrys Jones *Waters Corporation*	⊙ Determination of Aldehydes, Ketones, Short Chain Fatty Acids and Alkanes in Breath and Biofluids with GcxGC-TOF-MS-FID Antonis Myridakis *Imperial College London* Young Investigator Grant
1355 1415	⊙ Nontargeted Metabolite Profiling of Human Lung Epithelial Cells (A549) with HILIC Mode UPLC HRMS: Silica Nanoparticle Mediated Cytotoxicity Effects Kalyan Paila *Pemiscot Memorial Hospital*	... Keynote ...	... Extended Session ...	⊙ C-20 Hydrogenation is Involved in Intrauterine Cortisol Homeostasis Offering an Additional Sex-Driven Protection to Prenatal Stress. Implications in Birthweight Alex Gomez *Fundació Institut Mar d'Investigacions Mèdiques* Young Investigator Grant	⊙ Strategy for the Prediction of Neoadjuvant Therapy Response in Breast Cancer by MALDI-MSI Tiffany Porta *M4I institute, Maastricht University*	⊙ Deep Urinary Volatile Organic Compound Profiling with Headspace Sorptive Extraction and GCxGC-MS for Oesophago-Gastric Cancer Detection Qing Wen *Imperial College London* Young Investigator Grant

Scientific Session 7

Time	Mozart 1-3	Mozart 4-5	Papageno Hall	Paracelsus Hall	Trakl Hall	Doppler Hall
	Exploratory Metabolomics *Chair: Zdenek Spacil*	**Proteomics: Emerging Clinical Applications** *Chair: Renee Ruhaak* *2nd Chair: Tirsa van Duijl*	**Practical Training: Biomarker Discovery and Translational Research** *Chair: Grace van der Gugten*	**Small Molecule Keynote** *Chair: Ido Kema*	**MS Imaging: Clinical Diagnostics** *Chair: Tiffany Porta*	**Microbiology: Novel MS Clinical Applications** *Chair: Simon Cameron* *2nd Chair: Belen Rodriguez-Sanchez*
1430 1450	Untargeted Steroidomics for the Identification of Novel Steroid Profiles in Dysregulated Steroidogenesis Paal William Wallace *Technische Universität Dresden* Young Investigator Grant	The End of Kozak Dogma: The Ghost Proteome Michel Salzet *Université Lille*	Biomarker Discovery and Translational Research Leading to Clinical Utility: Experimental Approaches and Pitfalls Christiane Auray-Blais & Michel Boutin *Université de Sherbrooke*	Impact of the New European IVD Regulation on Medical Laboratories - Opportunities and Challenges Folker Spitzenberger *University of Applied Sciences Luebeck*	High-Throughput Analysis on FFPE Breast Cancer TMAs with DESI-MSI Olof Gerdur Isberg *University of Iceland* Young Investigator Grant	MBT FAST – Rapid Antimicrobial Susceptibility Testing by MALDI-TOF MS Markus Kostrzewa *Bruker Daltonik GmbH*
1450 1510	Untargeted Metabolomic Profiles of Newborns that Will or Will Not Develop Autism: A Pilot Study on Dried Blood Spots Julie Courraud *Statens Serum Institut*	Shortgun Proteomics for Differential Diagnosis of HPV-associated Cervix Transformation Natalia Starodubtseva *National Medical Research Center of Obstetrics*	... Extended Session ...	... Keynote ...	Laser Desorption Imaging – Rapid Evaporative Ionisation Mass Spectrometry Platform for Tissue Imaging Daniel Simon *Imperial College London* Young Investigator Grant	Detection of Vancomycin-Resistant Enterococcus faecium Using MALDI-TOF Mass Spectrometry Belén Rodríguez-Sanchez *Hospital General Universitario Gregorio Marañón*
1510 1530	Exploring the Volatomic Fingerprinting of Breast Cancer Tissue as an Untargeted Approach to Identify Potential Biomarkers Rosa Perestrelo *Centro de Química da Madeira, UMa* Young Investigator Grant	Multiplex Assay for Absolute Quantification of Inflammatory Proteins in 134 Meconium and First Stool Swabs from CELSPAC-TNG Birth Cohort Study Eliška Stuchlíková *Masaryk University* Young Investigator Grant	... Extended Session ...	... Keynote ...	Towards In vivo Molecular Diagnostics of Esogastric Cancer with Spidermass Real-Time, Mini Invasive Analysis Nina Ogrinc *University of Lille, PRISM laboratory* Young Investigator Grant	Development of Laser Assisted Rapid Evaporative Ionization Mass Spectrometry (LA-REIMS) as a Platform for Clinical Microbiology: The Past and the Future Simon Cameron *Queen's University Belfast*
1530 1545	**Intermission** *@ Entrance Foyer*					

MSACL-KSCC Satellite Meeting
Clinical Mass Spectrometry
May 24, 2020
Coex, Seoul, Korea
IFCC WorldLab SEOUL 2020

THURSDAY

	Scientific Session 8				
Mozart 1-3 **Metabolomics: Tackling Challenging Samples** *Chair: Elizabeth Want*	**Mozart 4-5** **MS Imaging Keynote** *Chair: Tiffany Porta*	**Papageno Hall** **Practical Training** *Chair: Grace van der Gugten*	**Paracelsus Hall** **Late-Breaking Science** *Chair: Jody van den Ouweland*	**Trakl Hall** **Lipidomics Keynote** *Chair: Anne Bendt*	**Doppler Hall**
1545 1605 ◎ Determination of D- and L-Lactic Acid in Urine by UPLC-MS with Electrospray Ionization Quantification Juliane Fagotti *Imperial College London* Young Investigator Grant	◎ Imaging the Unimaginable with Imaging Mass Cytometry Frits Koning *LUMC*	◎ Design of Experiments (DoE) – Get it Right from the Beginning Margrét Thorsteinsdottir *University of Iceland*	In-Matrix Derivitization of Indole Measurements by LC-MSMS Martin Faassen *University Medical Center, Groningen, the Netherlands*	◎ Skin Lipidomics in the Diagnosis and Treatment of Cutaneous Inflammation Anna Nicolaou *The University of Manchester*	
1605 1625 ◎ Optimising Laser Assisted - Rapid Evaporative Ionization - Mass Spectrometry Imaging (LA-REI-MSI) for the Spatially Resolved Analysis of Faecal Metabolites Petra Paizs *Imperial College London* Young Investigator Grant	... Keynote ...	... Extended Session ...	TBA	... Keynote ...	Track Closed
1625 1645 ◎ Steroidomics Profile Analysis by LCHR-MS in Human Seminal Fluid Eulalia Olesti *University of Geneve* Young Investigator Grant	... Keynote ...	... Extended Session ...	TBA	... Keynote ...	

1645
1700
Plenary Reception
@ *Entrance Foyer*
Ground Floor Bar to serve wine and other beverages for intermission while preparing for the poster awards.

1700
1715
Poster Award Presentation
@ *Mozart 4-5*
Chairs: Ilaria Belluomo and Julien Boccard

Closing Plenary Lecture
@ *Mozart 4-5*
Chair : Stephen Master

1715
1800

Donor-derived Cell-free DNA as a Biomarker in Organ Transplantation
Michael Oellerich
Institute for Clinical Pharmacology, University Medicine Göttingen, Germany

Molecular biomarkers have attracted special attention in solid organ transplantation because of unresolved problems that limit long-term outcome. There is a lack of reliable noninvasive markers. Immunosuppressive drug monitoring mainly indicates potential toxicity, but is a poor biomarker of graft damage. In kidney transplant patients, for example, an increase of plasma creatinine may be also be due to exsiccation, the use of ACE inhibitors, or immunosuppressive drug toxicity. By the time a rejection-related increase in plasma creatinine is evident, a significant degree of tissue damage has already occurred within the kidney. A further limitation of the current standard of care is that rejection episodes can only be confirmed by biopsies. Biomarkers are needed to achieve personalized immunosuppression to reduce premature graft loss. Against this background, a particularly promising new approach for the early detection of acute or chronic rejection or asymptomatic graft injury leading to irreversible damage is based on the determination of donor-derived circulating cell-free DNA (dd-cfDNA). Data on clinical validity have been documented in more than 48 independent studies which have shown that dd-cfDNA detects rejection episodes early, at an actionable stage, and is a more reliable marker of graft injury, compared to conventional tests. dd-cfDNA may also be useful to guide changes in immunosuppression, to monitor immunosuppression minimization (e.g. during tapering), and to prevent immune activation. The high negative predictive value of dd-cfDNA is the reason why this test can be helpful to avoid unnecessary biopsies. It could be shown that dd-cfDNA can be useful to detect subclinical (e.g. clinically unsuspected) graft damage as a result of immune activation triggered by under-immunosuppression. Early diagnosis of subclinical antibody-mediated rejection may improve outcomes after kidney transplantation. In summary, dd-cfDNA monitoring will allow more personalized treatment that shifts emphasis from reaction to prevention.

1800
1815
Final Remarks
@ *Mozart 4-5*
Chair: Elizabeth Want

1815
1900
Closing Petite Reception
@ *Entrance Foyer*
Wine & Cheese, Beer, light snacks. Winning posters to be on display

1900
CONGRESS CLOSED
@ *Entrance Foyer*

1900
2200
After Congress Meet-Up
@ *Your Choice*
Meet up at the Augustiner Braustubl (Historic Abby turned beer garden with food stalls) for an unofficial and unsponsored post-con gathering.

Clinical Mass Spectrometry   Education • Networking • Mentoring
MSACL 2021 EU
7th European Congress & Exhibits
Salzburg, Austria
September 14-16
Short Courses
September 12-14

Posters to be Presented

All posters (n=153) are to be placed by Tuesday at 16:30 and will remain placed for the duration of the congress. Poster removal is from 1300-1315 on Thursday.

Breath and VOC Analysis

◎ Poster #3c : Thursday : 09:30, attended for 1 hour
SHS-GC-MS Optimization for Volatile Organic Compounds Analysis in Oral Fluid and Urine
Bruno Costa, *University of São Paulo*

◎ Poster #4a : Wednesday : 10:00, attended for 1 hour
Impact of Oral Cleansing Strategies on Exhaled Volatile Organic Compound Levels
Bhamini Vadhwana, *Imperial College London*
Young Investigator Grantee

◎ Poster #5a : Wednesday : 10:00, attended for 1 hour
Developing a New Diagnostic Tool for Small Intestinal Bacterial Overgrowth following Upper-Gastrointestinal Surgery
Lory Hage, *Imperial College London*
Young Investigator Grantee

◎ Poster #10b : Wednesday : 15:30, attended for 1 hour
Evaluating the Stability of Breath During Storage in Thermal Desorption Tubes by Mass Spectrometry
Oscar Ayrton, *Imperial College London*
Young Investigator Grantee

◎ Poster #12b : Wednesday : 15:30, attended for 1 hour
Mass Spectrometry Techniques for Volatile Organic Compound Measurement in Cell and Organoid Media
Ilaria Belluomo, *Imperial College London*
Young Investigator Grantee

◎ Poster #13c : Thursday : 09:30, attended for 1 hour
Multi-Platform Correlation of Exhaled Volatile Organic Compounds by GC-MS, PTR-MS and SIFT-MS
Gengping Lin, *Imperial College London*
Young Investigator Grantee

◎ Poster #14c : Thursday : 09:30, attended for 1 hour
The Effect of Concentrations of Major Metabolites on Quantification of Minor Volatile Compounds in Breath by SESI-MS
Kseniya Dryahina, *J. Heyrovsky Institute of Physical Chemistry*

◎ Poster #16d : Thursday : 12:00, attended for 1 hour
Analyses of Volatile Organic Acids in Exhaled Breath Using Secondary Electrospray Ionization Mass Spectrometry (SESI-MS)
Suman Som, *J. Heyrovsky Institute of Physical Chemistry*

◎ Poster #18d : Thursday : 12:00, attended for 1 hour
Analysis of Human Monocyte Volatiles by Solid-Phase Microextraction Combined with Two-Dimensional Gas Chromatography-Mass Spectrometry for Biomarker Discovery
Antonio Pompeiano, *Central European Institute of Technology*

Data Science

◎ Poster #8a : Wednesday : 10:00, attended for 1 hour
Modelling the Within- and Between-run Variation in Internal Standard Signal to More Accurately Assess Ion Suppression Effects
Edmund Wilkes, *Imperial College Healthcare NHS Trust*
Young Investigator Grantee

◎ Poster #19a : Wednesday : 10:00, attended for 1 hour
Bioassay Classification Study via LC-MS and Machine Learning in Conjunction with Dimensionality Reduction
Ivan Plyushchenko, *Lomonosov Moscow State University*
Young Investigator Grantee

◎ Poster #26e : Wednesday : 10:00, attended for 1 hour
Comparison of Three Automated Data Processing and Reporting Approaches for Inborn Errors of Metabolism by LC-MS Flow Injection Analysis for Clinical Research
Shen Luan, *Thermo Fisher Scentific*

Endocrinology

◎ Poster #2d : Thursday : 12:00, attended for 1 hour
Confident Quantification of Steroids: Analysis in Human Plasma or Serum by LC-MS/MS for Clinical Research
Claudio De Nardi, *Thermo Fisher Scientific*

◎ Poster #7d : Thursday : 12:00, attended for 1 hour
Analysis of Aldosterone by LC-MS/MS: Comparison between a Conventional Electrospray Ionization Source and a New Atmospheric Pressure Ionization
Elisa Danese, *University of Verona*

◎ Poster #11a : Wednesday : 10:00, attended for 1 hour
Measurement of Oestrone and Oestradiol by Liquid-Chromatography Tandem Mass Spectrometry
Susan Johnston, *Department of Clinical Biochemistry*
Young Investigator Grantee

◎ Poster #14g : Thursday : 09:30, attended for 1 hour
Non-classic Congenital Adrenal Hyperplasia and Polycystic Ovary Syndrome Differentiation Using a Commercially Available LCMS/MS Steroid Profile Kit: A Case Report
Ivana Marković, *Osijek University Hospital*

◎ Poster #15c : Thursday : 09:30, attended for 1 hour
Analysis of Estrone and Estradiol to Low pg/mL Levels in Human Serum by Triple Quadrupole Mass Spectrometry for Clinical Research
Zuzana Skrabakova, *Thermo Fisher Scientific*

◎ Poster #21b : Wednesday : 15:30, attended for 1 hour
A Multi-Assay for the Quantification of Seven Steroid Hormones and Precursors in Serum with LC-MS/MS
Anna Lindahl, *Karolinska University Hospital*

◎ Poster #23a : Wednesday : 10:00, attended for 1 hour
Quantitation of 17β-Estradiol and Estrone in Serum by LC-MS/MS
Geoffrey Rule, *Millipore Sigma*

◎ Poster #24d : Thursday : 12:00, attended for 1 hour
A Sensitive Microflow LC-MS/MS Method for the Analysis of Corticosteroids in Human Plasma
Daniel Blake, *Sciex*

◎ Poster #25a : Wednesday : 10:00, attended for 1 hour
Development and Validation of an LC-MS/MS Method for the Measurement of Plasma Tryptophan, Kynurenine, Serotonin and 5-Hydroxyindoleacetic Acid
Leanne Sheppard, *Liverpool Clinical Laboratories*
Young Investigator Grantee

◎ Poster #25b : Wednesday : 15:30, attended for 1 hour
Simultaneous Determination of Cortisol and Cortisone in Saliva by LC-MS/MS: Method Validation
Marco Bagnati, *Maggiore della Carità Hospital*

◎ Poster #26g : Thursday : 09:30, attended for 1 hour
Evaluation of the Updated 25-Hydroxy Vitamin D Assay on the Cascadion SM Clinical Analyzer
Jenni Niemi, *Thermo Fisher Scientific Oy*

◎ Poster #27c : Thursday : 09:30, attended for 1 hour
Development of High Sensitivity Measurement of Arginine Vasopressin in Human Plasma by LC-MS/MS
Masaki Takiwaki, *Chiba University Hospital*

◎ Poster #27f : Wednesday : 15:30, attended for 1 hour
Rapid UPLC-MS/MS Dried Blood Spot Analysis of Steroid Hormones for Clinical Research
Dominic Foley, *Waters Corporation*

◎ Poster #27h : Thursday : 12:00, attended for 1 hour
A Liquid Chromatography Tandem-Mass Spectrometry Method for Determination of 9 steroids in Human Serum for Diagnosis of CAH and Steroidogenic Biosynthesis Defects
Xiaolin Wu, *Canterbury Health Laboratory*

◎ Poster #28a : Wednesday : 10:00, attended for 1 hour
Accuracy and Variability of Serum Bioavailable Testosterone Testing Methods
Yusheng Zhu, *Penn State University*

◎ Poster #28c : Thursday : 09:30, attended for 1 hour
Ultra-low Level Analysis of Aldosterone in Plasma Using the Xevo TQ-XS for Clinical Research
Ben Dugas, *Waters Corporation*

◎ Poster #20j : Wednesday : 15:30, attended for 1 hour
Attempt to Develop a Fast LC-MS/MS Application for the Simultaneous Measurement of 12 Biogenic Monoamines
Marcel van Borren, *Rijnstate Hospital*

◎ Poster #22e : Wednesday : 10:00, attended for 1 hour
Diagnosis of Carcinoid Tumors in the Small Intestine by LC-MS/MS Analysis of 5-Hydroxyindoleacetic Acid (5-HIAA) in Serum
Anders Lindgren, *Karolinska University Laboratory*

◎ Poster #23e : Wednesday : 10:00, attended for 1 hour
LC-MS/MS Analysis of Plasma Metanephrines Free of Interference from Midodrine and Metformin
Matthew Whitlock, *North West London Pathology*
Young Investigator Grantee

◎ Poster #23h : Thursday : 12:00, attended for 1 hour
Simultaneous Measurement of 18 Steroids in Human or Mouse Serum by LC-MS/MS to Profile the Classical and Alternate Pathways of Androgen Synthesis
Reena Desai, *University of Sydney*

Glycomics

◎ Poster #6a : Wednesday : 10:00, attended for 1 hour
Site-Specific Haptoglobin N-Glycosylation Changes in Colorectal Cancer
Jelena Simunovic, *Genos Glycoscience Research Laboratory*

◎ Poster #12a : Wednesday : 10:00, attended for 1 hour
Affinity Capture of a (Glyco)protein: Getting Your Sample Ready for the Mass Spectrometer
Valeriia Kuzyk, *VU Amsterdam/LUMC*
Young Investigator Grantee

◎ Poster #13e : Wednesday : 10:00, attended for 1 hour
Analysis of the Site-Specific N-Glycosylation of HeLa Cell Lysate
Simon Sugár, *Hungarian Academy of Sciences*
Young Investigator Grantee

Lipidomics

◎ Poster #3a : Wednesday : 10:00, attended for 1 hour
Implementation of Lipidomics in the Clinical Routine: Can Fluoride/Citrate Blood Sampling Tubes Improve Preanalytic Stability?
Lisa Hahnefeld, *Clinical Pharmacology, Goethe University Frankfurt*

◎ Poster #4c : Thursday : 09:30, attended for 1 hour
Lipidomic Profiling of Colorectal Cancer Cell Lines under Hypoxia
Juan Carlos Alarcon-Barrera, *LUMC*
Young Investigator Grantee

◎ Poster #5c : Thursday : 09:30, attended for 1 hour
Determination of Membrane Lipid Gangliosides in Human Serum and Cerebral Organoids
Gabriela Dovrtělová, *Recetox centre, Masaryk University*
Young Investigator Grantee

◎ Poster #9b : Wednesday : 15:30, attended for 1 hour
Development and Validation of a LC-MS/MS-based Assay for Quantification of Polyunsaturated Fatty Acids from Human Plasma and Red Blood Cells
Vlad Serafim, *'Victor Babeş' University of Medicine and Pharmacy*

◎ Poster #15d : Thursday : 12:00, attended for 1 hour
Lipid Composition of Cancer-Derived Extracellular Vesicles and its Potential for the Identification of Body Fluid-Based Biomarkers for Breast Cancer diagnosis
Erika Dorado, *Imperial College London*
Young Investigator Grantee

◎ Poster #21c : Thursday : 09:30, attended for 1 hour
Graft Quality Assessment in Kidney Transplantation by Monitoring Lipidomic Changes in the Organ During Transplantation Using Solid Phase Microextraction (SPME)
Natalia Warmuzińska, *Nicolaus Copernicus University in Toruń*
Young Investigator Grantee

◎ Poster #23b : Wednesday : 15:30, attended for 1 hour
Plasma Lysosphingolipid Analysis by LC-Differential Mobility Spectrometry-MS/MS (LC-DMS-MS/MS): Rapid Sample Preparation and Resolution of Stereoisomers
Cyrus Papan, *SCIEX*

Metabolites & Metabolomics

◎ Poster #1d : Thursday : 12:00, attended for 1 hour
Targeted Metabolomics Reveals an Influence of the FTO Gene on the Kynurenine Pathway
La-or Chailurkit, *Mahidol University*

◉ Poster #2a : Wednesday : 10:00, attended for 1 hour
Enhanced Mass Spectrometric Profiling of the Human Blood Exposome Using an Optimised Dispersive Solid Phase Extraction Protocol
Mark David, *University of New South Wales*
Young Investigator Grantee

◎ Poster #2c : Thursday : 09:30, attended for 1 hour
Detection of F-2 Mycotoxin Zearalenone in Human Urine. Consumption of Contaminated Cereal Crops or Doping Offence?
George Madalin Danila, *Romanian Doping Control Laboratory*
Young Investigator Grantee

◉ Poster #5b : Wednesday : 15:30, attended for 1 hour
Prostate Cancer Metabolic Alterations Induced by Zika Virus
Jeany Delafiori, *University of Campinas*
Young Investigator Grantee

◎ Poster #6b : Wednesday : 15:30, attended for 1 hour
Biomarker Analysis by Mass Spectrometry and Artificial Intelligence Techniques of Obese Human Plasma
Flávia Luísa Dias Audibert, *University of Campinas*
Young Investigator Grantee

◎ Poster #9a : Wednesday : 10:00, attended for 1 hour
Simple and Rapid Tandem Mass Spectrometry Method for the Analysis of Methylmalonic Acid in Urine
Michel Boutin, *Université de Sherbrooke*

◎ Poster #9d : Thursday : 12:00, attended for 1 hour
Serum Biomarkers of Chemoradiosensitivity in Esophageal Cancer is Identified by the Targeted Metabolomics Approach
Masaru Yoshida, *Kobe University Graduate School of Medicine*

◎ Poster #10c : Thursday : 09:30, attended for 1 hour
Method Development of Amino Acid Analysis in a Dried Blood Spot for the Second-Tier Test in the Newborn Screening Program
Zahra Talebpour, *Department of chemistry, Alzahra University*

◉ Poster #11b : Wednesday : 15:30, attended for 1 hour
Metabolomic Barometer of Gestational and Postpartum Weight in Overweight Pregnant Women
Chung-Ho E. Lau, *Imperial College London*
Young Investigator Grantee

◎ Poster #13d : Thursday : 12:00, attended for 1 hour
Establishment of a Liquid-Chromatography Tandem Mass-Spectrometry Method for Vitamin D Metabolites to Detect 24-Hydroxylase Deficiency
Sieglinde Zelzer, *Medical University Graz*

◎ Poster #15b : Wednesday : 15:30, attended for 1 hour
'Functional Microbiomics' – Standardized Assessment of Nutrition-Microbiome-Host Interplay by Targeted Metabolomics
Therese Koal, *BIOCRATES Life Sciences AG*

◎ Poster #17a : Wednesday : 10:00, attended for 1 hour
GC-MS Determination of Candidate Target Biomarkers for Early Detection of Oesophageal Squamous Cell Cancer
Yan Mei Goh, *Imperial College London*
Young Investigator Grantee

◎ Poster #17c : Thursday : 09:30, attended for 1 hour
Different Approaches for Vitamin D Determination in Newborns by LC-MS/MS
Rafal Rola, *Nicolaus Copernicus University*
Young Investigator Grantee

◎ Poster #18c : Thursday : 09:30, attended for 1 hour
Maternal Glutaric Aciduria Type 1 (GA 1) Detected Through Newborn Screening in Croatia
Ivana Križić, *University Hospital Center Zagreb*

◎ Poster #19e : Wednesday : 10:00, attended for 1 hour
B Vitamin Reference Ranges Determination Using HPLC-MS/MS and Retrospective Statistical Analysis
Arseny Sadykov, *Scientific laboratory "Chromolab", Moscow, Russia*
Young Investigator Grantee

◎ Poster #20a : Wednesday : 10:00, attended for 1 hour
Comprehensive Clinical Acylcarnitines by LC-MS/MS
Steve Bruce, *University Hospital of Lausanne (CHUV)*

◉ Poster #20c : Thursday : 09:30, attended for 1 hour
Solid Phase Microextraction (SPME) in Kidney Examination – LC-MS/MS-based Identification of Potentially Significant Metabolites in Graft Quality Assessment
Iga Stryjak, *Nicolaus Copernicus University in Toruń*
Young Investigator Grantee

◉ Poster #22c : Thursday : 09:30, attended for 1 hour
Analysis of Changes in Bile Acids Concentration in Bile in Response to the Degree of Liver Ischemia and the Method of Organ Preservation
Kamil Łuczykowski, *Nicolaus Copernicus University in Toruń*
Young Investigator Grantee

◎ Poster #26a : Wednesday : 10:00, attended for 1 hour
Pharmacometabolomic Study of Novel Multitarget Drugs Based on Natural Prostaglandins in Terms of Therapeutic Effectiveness
Kseniia Shestakova, *University of Verona, Sechenov University*

◎ Poster #26c : Thursday : 09:30, attended for 1 hour
Evaluation of an Artificial Serum as a Surrogate Matrix for Calibration Samples for a Preeclampsia Risk Prediction Test
Anna Catharina Suhr, *Metabolomic Diagnostics*

◎ Poster #26f : Wednesday : 15:30, attended for 1 hour
Advancement of the Quantitative Measurement of Enzyme Activities in Six Lysosomal Storage Disorders via LC-MSMS
Mietha Magdalena van der Walt, *Human Metabolomics, North-West University, SA*

Poster #27b : Wednesday : 15:30, attended for 1 hour
Comparison of Phenylketonuria Screening with a Fluorimetric Method and with Tandem Mass Spectrometry
Vanja Cuk, *University Children's Hospital*
Young Investigator Grantee

Poster #27d : Thursday : 12:00, attended for 1 hour
Untargeted Metabolomic Profiling of Plasma Samples of Patients with MBOAT7 Gene Defect
Basri Gülbakan, *Hacettepe University*

Poster #27g : Thursday : 09:30, attended for 1 hour
Tandem Mass Spectrometry-based Analysis Reveal Relationship between Active DNA Demethylation and Krebs Cycle in AML and MDS
Justyna Szpotan, *Collegium Medicum in Bydgoszcz*
Young Investigator Grantee

Poster #28f : Wednesday : 15:30, attended for 1 hour
Simultaneous Quantitation of Diabetes Markers and Comprehensive Metabolome Annotation Achieved via Semi-targeted Analysis of Serum Samples
Ioanna Ntai, *Thermo Fisher Scientific*

Poster #20k : Thursday : 09:30, attended for 1 hour
Serum Metabolomics Using LC-MS Reveals Potential Biomarker of Myocardial Ischemia
Sunhee Jung, *Korea Basic Science Institute*

Microbiology

Poster #7b : Wednesday : 15:30, attended for 1 hour
Enterobacter: When MALDI-TOF (VITEK® MS) and New Taxonomy Speak as One Voice
Valérie Monnin, *bioMérieux*

Poster #11f : Wednesday : 15:30, attended for 1 hour
Improvement of the Experimental and Informatic Pipeline for High-Throughput MS/MS Proteotyping of Pathogens
Karim Hayoun, *CEA*

Poster #16b : Wednesday : 15:30, attended for 1 hour
Monitoring of Human Procalcitonin by Functionalized MALDI Surfaces
Josef Dvorak, *Department of Biochemistry, Charles University*

Poster #19c : Thursday : 09:30, attended for 1 hour
Detection of Carbapenemase-producing Organisms via MALDI-TOF
Narae Hwang, *Kyungpook National University*

Proteins & Proteomics

Poster #2b : Wednesday : 15:30, attended for 1 hour
Semi-Automated Positive Pressure SPE for Proteomics
Christian Scherling, *Tecan*

Poster #9c : Thursday : 09:30, attended for 1 hour
Mass Spectrometry Studies of Glutathione S Transferase P1 Protein and Gene and 5-Methylcytosine Level in Acute Lymphoblastic Leukemia
Alireza Ghassempour, *Shahid Beheshti University*

Poster #11d : Thursday : 12:00, attended for 1 hour
Developing the Research to Routine Workflows with FAIMS: Automating Large-Scale SRM Method Creation for Routine Plasma Proteomics Screening
Paul Humphrey, *ThermoFisher Scientific*

Poster #12d : Thursday : 12:00, attended for 1 hour
Separation of Intact Parathyroid Hormone and Variants Using a Highly Sensitive Sheathless CE-ESI-MS/MS Method
Laurent Nyssen, *University of Liège*

Poster #13a : Wednesday : 10:00, attended for 1 hour
Quantitative Determination of Human IgA Subclasses and their Fc-Glycosylation Patterns in Human Plasma by Using Peptide Analogue Internal Standard and UHPLC-MS/MS
Hsiao-Fan Chen, *Taipei Medical University*
Young Investigator Grantee

Poster #14b : Wednesday : 15:30, attended for 1 hour
LC-MS-based Pharmacokinetics of Insulin Aspart Measured in Individuals with Type 2-Diabetes During Operation Closed-Loop Glucose Control
Jean-Christophe Prost, *Bern University Hospital*

Poster #15a : Wednesday : 10:00, attended for 1 hour
Confident Protein and Peptide Identification with Mass Spectrometry: Understanding and Exploiting Collision Energy Dependence
Agnes Revesz, *Research Centre for Natural Sciences, HAS*
Young Investigator Grantee

Poster #16c : Thursday : 09:30, attended for 1 hour
Expression of Cellular and Neurodevelopmental Markers in Cerebral Organoids
Marketa Nezvedova, *Masaryk University*

Poster #17b : Wednesday : 15:30, attended for 1 hour
Peptide Sequencing with Top-Down Synthesized TiO2 Nanowires Using Laser Desorption/Ionization Time-of-Flight Mass Spectrometry
Tae Gyeong Yun, *Yonsei University*
Young Investigator Grantee

Poster #17d : Thursday : 12:00, attended for 1 hour
Prognostic Potential of Combined Omics Analysis for Diagnostics of Azoospermia
Alexnder Brzhozovskiy, *National Medical Research Center for Obstetrics*
Young Investigator Grantee

Poster #18a : Wednesday : 10:00, attended for 1 hour
Single-Shot Targeted Proteomics Flavivirus Assay
Jayantha Gunaratne, *Institute of Molecular and Cell Biology, A*STAR*

Poster #20d : Thursday : 12:00, attended for 1 hour
Development of a Sensitive Mass Spectrometric Method for the Quantification of Procalcitonin
Sebastian-Alexander Tölke, *Hochschule Fresenius, Germany*

Poster #20e : Wednesday : 10:00, attended for 1 hour
Exploring Optimal Protein Identification and Repeatability via Sequences of Fine-Tuned LC-MS/MS Runs
Márton Gyula Milley, *RCNS, Hungarian Academy of Sciences*
Young Investigator Grantee

Poster #22d : Thursday : 12:00, attended for 1 hour
Data-Independent Acquisition Spiked with QconCAT Quantification Standards (DISQ):An Alternative to Targeted Methods?
Florian Christoph Sigloch, *PolyQuant GmbH*

◎ Poster #25c : Thursday : 09:30, attended for 1 hour
Development and Comparison of Two High-throughput LC-MS Methods for the Accurate Quantitation of IGF1 in Human Serum
Anders Fridström, *Merck Sweden AB*

◎ Poster #26b : Wednesday : 15:30, attended for 1 hour
High-Field Asymmetric Waveform Ion Mobility Spectrometry (FAIMS) - Parallel Reaction Monitoring (PRM) for HER2 Protein Quantitation in Tumor Biopsies
Steve Sweet, *AstraZeneca*

◎ Poster #27a : Wednesday : 10:00, attended for 1 hour
Characterizing the Novel Metastatic Regulator in CRPC-based on Quantitative Proteomics
Do Eun Kim, *College of Pharmacy, Kyungpook National University*

◎ Poster #27e : Wednesday : 10:00, attended for 1 hour
SCOUT Rtbeads, A Useful Tool for LC-MS Retention Time Control: A Relevant Application for Multiplex LC SRM Biomarker Quantitation
Chloe Bardet, *ANAQUANT*

◎ Poster #28b : Wednesday : 15:30, attended for 1 hour
Incorporating Stable Isotope-Labeled IgG Internal Standard and Affinity Purification for Analyzing Human IgG4 and Fc-glycan Profiles by UHPLC-MS/MS
Isabel I-Lin Tsai, *Taipei Medical University*

◎ Poster #28d : Thursday : 12:00, attended for 1 hour
Direct Monitoring of Fucosylated Glycopeptides of Alpha-Fetoprotein in Hepatocellular Carcinoma Serum by LC-MS/MS with Immunoprecipitation
Kwang Hoe Kim, *Korea Basic Science Institute*

◎ Poster #20m : Thursday : 12:00, attended for 1 hour
High-Throughput PEAKS Workflow and SPIDER Algorithm for Large Scale Clinical Proteomics and Variant Identification Using PEAKS Online X
Jonathan Krieger, *Bioinformatics Solutions Inc*

Quality Management & Standardization

◎ Poster #4b : Wednesday : 15:30, attended for 1 hour
Development of a Candidate Higher Order Reference Measurement Procedure for the Analysis of Metanephrines in Plasma Using Micro-Flow LC-MS/MS
Dima AlMekdad, *King's College London*
Young Investigator Grantee

◎ Poster #10a : Wednesday : 10:00, attended for 1 hour
Intact-Cell Mass Spectrometry for Monitoring of Stem Cells Cultures
Petr Vanhara, *St. Anne's University Hospital Brno, Czech Republic*

◎ Poster #11e : Wednesday : 10:00, attended for 1 hour
Methodological Traceability: Standardized Description of LC-MS-based Measurement Methods in Laboratory Medicine
Carina Schuster, *LMU Munich*

◎ Poster #21e : Wednesday : 10:00, attended for 1 hour
Collision-energy breakdown curves – an additional tool to characterize MS/MS methods
Sophie Mörlein, *University Hospital, LMU Munich*

Small Molecules / Tox / TDM

◎ Poster #1a : Wednesday : 10:00, attended for 1 hour
The Effect of Sample Tube Type, pH, Storage Time and Temperature on Antihypertensive Non-Adherence Results by Quantitative LC-MS/MS
Aneela Iqbal, *Leicester Royal Infirmary*
Young Investigator Grantee

◎ Poster #1c : Thursday : 09:30, attended for 1 hour
Influence of Saw Palmetto and Pygeum Africana Extracts on the Urinary Concentration of Anabolic Androgenic Steroids
Francesco Botre, *"Sapienza" University of Rome & Antidoping Lab FMS*

◎ Poster #3b : Wednesday : 15:30, attended for 1 hour
Simultaneous Determination of Vitamins B1, B2 and B6 in Whole Blood by LC-MSMS
Stefanie Wernisch, *Waters GmbH*

◎ Poster #5d : Thursday : 12:00, attended for 1 hour
Simple and High-throughput LC-MS/MS Method for Simultaneous Measurement of Five Smoking-related Metabolites in Urine
Mi ryung Chun, *Samsung Medical Center*

◎ Poster #6c : Thursday : 09:30, attended for 1 hour
High-throughput Quantification of Immunosuppressant Drugs in Human Blood by LC-MSMS for Clinical Research
Magnus Olin, *Thermo Fisher Scientific*

◎ Poster #6d : Thursday : 12:00, attended for 1 hour
Validated High-Sensitive Steroid Profiling in Human Serum Using LC-MS/MS
Stephane Moreau, *SFP*

◎ Poster #8b : Wednesday : 15:30, attended for 1 hour
Towards a LC-MS/MS Method for the Quantification of Serum Symmetric Dimethylarginine and Asymmetric Dimethylarginine for Use in a Routine Clinical Laboratory
David Marshall, *Manchester University NHS Foundation Trust*
Young Investigator Grantee

◎ Poster #8c : Thursday : 09:30, attended for 1 hour
Direct Quantification of Amino Acids in Human Plasma by LC-MS/MS
Valérie Thibert, *Thermo Scientific France*

◎ Poster #8d : Thursday : 12:00, attended for 1 hour
Simultaneous Determination of 14 Antiretroviral Drugs in Plasma by micro-LC-MS/MS for Therapeutic Drug Monitoring
Isabela Tarcomnicu, *National Institute of Infectious Diseases "Prof. Dr. Matei Bals"*

◎ Poster #10d : Thursday : 12:00, attended for 1 hour
Untargeted Data Independent Acquisition Analysis of Drugs of Abuse by LC-MS/MS, a Generic Approach for Routine Clinical Pathology Screening
Emily Armitage , *Shimadzu Corporation*

◎ Poster #11c : Thursday : 09:30, attended for 1 hour
Fast Multiplexed Analysis of Cannabinoids and their Metabolites in Urine Using MassHunter StreamSelect LC-MS System
Moritz Wagner, *Agilent Technologies*

◎ Poster #12c : Thursday : 09:30, attended for 1 hour
Development and Evaluation of a LC-MS/MS Method for the Quantitation of Isavuconazole in Human Serum Samples
Andreas Meinitzer, *Medical University Graz*

◎ Poster #13b : Wednesday : 15:30, attended for 1 hour
LC-MS/MS Method for Screening of Intoxication and Drug Adherence of Angiotensin Converting Enzyme Inhibitors in Plasma
Mohsin Ali, *Institute of Clinical Pharmacy and Pharmacotherapy*
Young Investigator Grantee

◎ Poster #14a : Wednesday : 10:00, attended for 1 hour
Implementation of Triple Quadrupole LC-MS/MS Complete Kits on a High Resolution Mass Spectrometer – Limitations and Benefits
Katharina Kern, *RECIPE Chemicals + Instruments GmbH*

◎ Poster #14d : Thursday : 12:00, attended for 1 hour
Fully-Automated LC-MS Method for Uracil and Dihydrouracil in Human Plasma
Doriane Toinon, *Shimadzu France*

◎ Poster #14f : Wednesday : 15:30, attended for 1 hour
A Fast and Robust Method for the Quantification of Venetoclax by LC-MS/MS
Regula Steiner, *University Hospital Zürich*

◎ Poster #14h : Thursday : 12:00, attended for 1 hour
A High-throughput LC-MS/MS Method for Vitamin A and Vitamin E
Helge Berland, *Hormone Laboratory, Haukeland University Hospital*

◎ Poster #18b : Wednesday : 15:30, attended for 1 hour
Simultaneous Determination of Antihypertensive Drugs in Serum by LC-MS/MS
Karolina Kotalova, *Spadia LAB a.s.*

◎ Poster #19b : Wednesday : 15:30, attended for 1 hour
Integrated Toxicological Screening and Confirmation Analysis with Stable 24/7 Availability
Frank Streit, *UMG Goettingen, Germany*

◎ Poster #19d : Thursday : 12:00, attended for 1 hour
Determination of Biomarkers of Organophosphorus Agent Intoxication by Ion Chromatography and Tandem Mass Spectrometry
Igor Rodin, *Lomonosov Moscow State University*

◎ Poster #20b : Wednesday : 15:30, attended for 1 hour
An LC-MS/MS Method for Simultaneous Measurement of Five Antimicrobial Drugs (Vancomycin, Gentamicin, Amikacin, Linezolid and Teicoplanin) for TDM
Antonio Martini, *ASST OVEST MILANESE;CLINICAL LAB. Legnano Hospital*

◎ Poster #20f : Wednesday : 15:30, attended for 1 hour
A UHPLC-MS/MS Method for the Simultaneous Quantification of Ten Directly Acting Anti-HCV Drugs in Patient Plasma
Amedeo De Nicolò, *University of Turin*

◎ Poster #20h : Thursday : 12:00, attended for 1 hour
Automated Sample Pretreatment for the Analysis of Synthetic Cannabinoids in Urine with LC-MS/MS
Stefan Eckelt, *Labor Berlin*

◎ Poster #20i : Wednesday : 10:00, attended for 1 hour
Drugs of Abuse Screening and Quantification Directly from Urine Using Paperspray Technology
Shawnna Donop, *ThermoFisher Scientific*

◎ Poster #21a : Wednesday : 10:00, attended for 1 hour
Simultaneous Quantitation of Controlled-Substances/Pain-Management Drugs in Oral Fluids via Coated Blade Spray-Tandem Mass Spectrometry
Frances Carroll, *Restek Corporation*

◎ Poster #22b : Wednesday : 15:30, attended for 1 hour
Phospholipid Removal from Protein Precipitated Plasma Using In-Line Sample Preparation (ILSP)
Christian Weyer, *Restek Corporation*

◎ Poster #23c : Thursday : 9:30, attended for 1 hour
Performance of Common Methods When Assessing Protein-Binding in Low-Protein and Low-Volume Samples
Reingard Raml, *Joanneum Research Institute*

◎ Poster #23d : Thursday : 12:00, attended for 1 hour
Evaluation of Simplified Workflows for Hair Matrix Extraction Prior to UHPLC-MS/MS Analysis
Alan Edgington, *Biotage GB Limited*

◎ Poster #25e : Wednesday : 10:00, attended for 1 hour
LC-MS/MS Analysis of Plasma Epinephrine and Norepinephrine
Marianne Bergmann, *Lillebaelt Hospital*

◎ Poster #26d : Thursday : 12:00, attended for 1 hour
Impact of Body Mass Index on Cytochrome P450 3A Phenotype in Obese Patients
Birgit M. Wollmann, *Diakonhjemmet Hospital, Oslo, Norway*

◎ Poster #28e : Wednesday : 10:00, attended for 1 hour
New Solutions Applied in Oral Fluid Drug Testing: Fine-Tuning and Optimization of the SPME-LC-MS Method
Łukasz Sobczak, *Dept of Pharmacodynamics and Molecular Pharmacology, Faculty of Pharmacy, Collegium Medicum in Bydgoszcz at Nicolaus Copemicus University in Toruń*
Young Investigator Grantee

◎ Poster #28g : Thursday : 09:30, attended for 1 hour
Reversed-phase UPLC-MS/MS Analysis of Plasma Catecholamines and Metanephrines for Clinical Research
Gareth Hammond, *Waters Corporation*

◎ Poster #28h : Thursday : 12:00, attended for 1 hour
Confirmation Method for the Determination of Drugs of Abuse in Saliva: Validation of a RUO Kit by LC-MS/MS, Ready for Use.
Ugo de Grazia, *Fondazione IRCCS Istituto Neurologico Carlo Besta*

◎ Poster #21f : Wednesday : 15:30, attended for 1 hour
Validation of a LCMS Assay for Citrulline Suitable for Daily Routine Through an Original MRM Approach
Claude Hercend, *Aphp Hupnvs, UF de Biochimie Clinique, Hopital Bichat*

◎ Poster #21h : Thursday : 12:00, attended for 1 hour
Development and Validation of the First UHPLC-MS/MS Method for the Quantification of the New Anti-Ebola Drug Remdesivir: Application to Healthy Volunteers
Valeria Antonio D, *University of Turin*

◎ Poster #22f : Wednesday : 15:30, attended for 1 hour
Developing an LC-MS/MS Method for 29 Antihypertensive Drugs in Urine
Melissa McNaughton, *NHS*

◎ Poster #23f : Wednesday : 15:30, attended for 1 hour
LC-MS/MS Measurement of Phosphatidylethanol (PEth 16:0/18:1) and Drugs of Abuse Applying Volumetric DBS Device
Gökçe Göksu Gürsu, *Sem Laboratuar Cihazları Pazarlama San. Ve Tic A.Ş*

◎ Poster #23g : Thursday : 09:30, attended for 1 hour
Performance of the Cascadion SM Immunosuppressants Panel on the Cascadion SM Clinical Analyzer
Mari Kiviluoma, *Thermo Fisher Scientific Oy*

Tissue Imaging

◎ Poster #1b : Wednesday : 15:30, attended for 1 hour
Brain Tumour Characterization Using Laser Desorption Imaging – Rapid Evaporative Ionisation Mass Spectrometry
Hanifa Koguna, *Imperial College London / NPL*
Young Investigator Grantee

◎ Poster #3d : Thursday : 12:00, attended for 1 hour
Probing Aβ Dynamics in APP Knock-In Mice and *in vitro* Using Stable Isotope Labelling and MALDI Imaging Mass Spectrometry, and Examining the Role of Microglia
Katie Stringer, *University College London*
Young Investigator Grantee

◎ Poster #4d : Thursday : 12:00, attended for 1 hour
Pyroglutamation of Amyloid-β-42 (Aβ1-42) Followed by Aβ1-40 Deposition Underlies Plaque Polymorphism in Progressing Alzheimer's Disease Pathology
Jörg Hanrieder, *University of Gothenburg*

◎ Poster #13f : Wednesday : 15:30, attended for 1 hour
MALDI-TOF MSI of MeLiM Melanoma: Searching for Differences in Protein Profiles
Lucie Vanickova, *Mendel University in Brno*

◎ Poster #21d : Thursday : 12:00, attended for 1 hour
Cross-modality Single-pixel Correlation of Multimodal Imaging Mass Spectrometry of Prostate Cancer
Ambra Dreos, *Sahlgrenka Academy at the University of Gothenburg*

◎ Poster #22a : Wednesday : 10:00, attended for 1 hour
Exploring Healthy and Tumor Tissue Microenvironment with Immuno-Oncology Markers Using Multiplexed Hyperion Imaging System
Roberto Spada, *Fluidigm*

◎ Poster #21g : Thursday : 09:30, attended for 1 hour
Mass Spectrometric Imaging of Cysteine Rich Proteins in Human Skin
Tomas Do, *Mendel University*

clinical mass spectrometry

cms

an international journal

Troubleshooting

◎ Poster #24a : Wednesday : 12:30, attended for 15 min
The Hurdles of Developing an LC-MS/MS Assay for Desmosine, a Biomarker for Elastin Degradation
Jody van den Ouweland, *Canisius-Wilhelmina Hospital*

◎ Poster #24b : Wednesday : 12:45, attended for 15 min
Fast and Efficient Method for Detection and Quantification of Catecholamines in Human Plasma by LC-MS/MS - Troubleshooting
Chiara Calaprice, *University hospital of Liege*
Young Investigator Grantee

◎ Poster #24c : Wednesday : 13:00, attended for 15 min
Matrix Dilemmas in LC-MS/MS: The Use of Appropriate Calibrators and Internal Standards to Facilitate Acceptable Criteria for Quantification
Tania Grobler, *North-West University*

Various Other

◎ Poster #7c : Thursday : 09:30, attended for 1 hour
Rapid Evaporative Ionisation Mass Spectrometry (REIMS): A Diagnostic Tool for Omental Metastases in Patients with Primary Ovarian Cancer
Eftychios Manoli, *Imperial College London*
Young Investigator Grantee

◎ Poster #11g : Thursday : 09:30, attended for 1 hour
Fully Automated LC-MS/MS Analysis of Anticoagulants Using a Novel Reagent Kit
Franck Chevallier, *Alsachim*

◎ Poster #14e : Wednesday : 10:00, attended for 1 hour
Confident Quantitation of 25-Hydroxyvitamin D2 and D3 in Human Plasma for Clinical Research by LC-MSMS
Mariana Barcenas, *Thermo Fisher Scientific*

◎ Poster #14i : Wednesday : 10:00, attended for 1 hour
Analysis and Characterization Modified *in vitro* Transcribed RNA Using LC-MS/MS Method
Dominika Strzelecka, *Faculty of Physics, University of Warsaw*

◎ Poster #16a : Wednesday : 10:00, attended for 1 hour
Prednisone and Prednisolone Detection in Urine Samples: A GC-C-IRMS Method to Discriminate Exogenous or Endogenous Origin
Loredana Iannella, *Laboratorio Antidoping, FMSI*

◎ Poster #19f : Wednesday : 15:30, attended for 1 hour
Targeted MMulti-OMICS: Rapid Plasma Profiling of a Bladder and Lung Cancer Human Cohort
Billy Molloy, *Waters*

◎ Poster #25d : Thursday : 12:00, attended for 1 hour
Integration of Mycophenolate and its Metabolite Analysis in Plasma Using LC-MS/MS with Full-Automated Sample Preparation
Fanny Dayot, *Alsachim*

◎ Poster #22g : Thursday : 09:30, attended for 1 hour
Validation of the Doubly-labelled Water Method to Determine Total Energy Expenditure in a Rat Model
Anita Eberl, *Joanneum Research*

Exhibitor Summaries

AFFINSEP (Mini-Table) : The art of Making Sample Preparation Easier.
http://www.affinisep.com
Discover the best Micro-Extraction solutions: Stage-tips, Spin-SPE, 96 well-plate ... for analysis of Proteins, Peptides, small molecules.

Agilent Technologies (Booth #19-20)
http://www.agilent.com
With an industry leading portfolio of analytical products for your clinical research laboratory, Agilent Technologies delivers everything your laboratory needs from sample preparation to final answers. From automation and sample preparation technologies, columns and consumables, laboratory informatics, to liquid and gas chromatography systems, and mass spectrometry systems such as ICP-MS, GC/MS, and LC/MS, Agilent provides premiere analytical solutions that will ensure confident identification and quantitation of both endogenous and exogenous substances in complex biological matrices with the utmost accuracy, productivity and reliability.

Biocrates Life Sciences (Mini-Table Row)
http://www.biocrates.com
Biocrates and Metanomics Health have recently merged to create a new global leader in Metabolomics and early disease detection. As a total solution provider, Biocrates and Metanomics Health provide the broadest technology and product portfolio in the industry. The merged companies have extensive expertise in targeted and global metabolic profiling, customized assay development, targeted screening kits, and data interpretation. Finally, capabilities in CDx / Diagnostic Kit development make Biocrates and Metanomics Health the preferred partner for all your requirements in metabolite analysis. Biocrates' metabolomics kits allow for the analysis of up to 400 metabolites across multiple analyte classes. These kits have contributed to more than 800 scientific publications and are being used in >100 mass spectrometry laboratories throughout the world. The whole range of analytical and data interpretation services is available through the company's service laboratories in Berlin, Germany and Innsbruck, Austria.

Bioinformatics Solutions (Booth #23)
http://www.bioinfor.com/
Bioinformatics Solutions Inc. (BSI) is the world leading solution provider for mass spectrometry-based proteomics and targeted protein studies. By collaborating with clinical laboratories, and biotechnology companies using mass spectrometry for clinical- and diagnostic-based applications, BSI aims to advance proteomics through professional data analysis software and service.

Biotage AB, Sweden (Booth #5)
http://www.biotage.com
Biotage AB, Sweden is a global leader in life science technology. With a broad scope of tools for synthesis, work-up, purification, evaporation and analysis, the company provides knowledge and expertise in the areas of analytical chemistry and medicinal chemistry.

Cambridge Isotope Labs (Booth #21)
http://www.isotope.com
Cambridge Isotope Laboratories, Inc. is the world leader in the manufacture and separation of stable isotopes and isotope-labeled compounds. CIL and Euriso-Top (a European subsidiary of CIL) offer highly pure compounds that are uniformly or selectively enriched in 13C, 15N, D, 18O or 17O. CIL's labeled reagents are used in proteomics, metabolomics, metabolism, and environmental applications for quantitative mass spectrometry. Our products include MRM PeptiQuantTM assay kits, SILAC protein quantitation kits, media and reagents, 99% enriched amino acids, Mouse Express® Lys 13C6 and 15N mouse feed and tissue, 15N spirulina, intact labeled proteins, growth media for protein expression, cell-free protein synthesis products, environmental contaminants standards for ultra-trace analysis, steroids, acylcarnitines, drug metabolites, nucleic acids, lipids and carbohydrates. CIL has cGMP capabilities; a majority of substrates can be manufactured to Q7A compliance.

Capitainer (Mini-Table Row)
https://capitainer.se/
Capitainer enables patients to collect their own volume-defined dried blood spot at home.

Chromsystems (Booth #8)
http://www.chromsystems.com
Chromsystems is a leading global company providing ready-to-use kits, multilevel calibrators and quality controls for routine clinical diagnostics by LC-MS/MS and HPLC. Our parameter menu covers a range of areas such as newborn screening, therapeutic drug monitoring, steroid analysis, vitamin profiling and more. We continuously expand our portfolio with additional tests all ensuring a highly accurate and cost-effective analysis. We enable laboratories to add new parameters into their diagnostic routine and expand their testing menu without prior technical expertise. They can immediately start the analysis with a minimum of time for the sample preparation. The products are comprehensively validated, and in particular LC-MS/MS methods with all widely used tandem mass spectrometers. They are CE-IVD compliant, satisfying regulatory requirements in the laboratory. We combine these high quality products with an excellent support programme and service for our customers.

Eureka Lab Division (Booth #10)
https://www.eurekaone.com/
Eureka Lab Division is a young company founded in the late 90s; it has identified its core business in the area of clinical toxicology . To perform this type of analysis is necessary to use standard analytical techniques such as LC, liquid chromatography, also called HPLC (high performance liquid chromatography), Gas Chromatography or LC coupled to a mass detector. The ONLY Italian Company that researches, develops and produces diagnostic KITS for HPLC, GC, GC-MS, LC-MS/MS.

Fluidigm (Mini-Table Row)
http://www.fluidigm.com
Fluidigm is committed to empowering the imaging community with research tools to deeply interrogate cell phenotypes and function. Using Fluidigm Imaging Mass Cytometry™ , you can obtain high-dimensional cellular phenotypes in the spatial context of the tissue microenvironment at single-cell resolution, measuring >40 parameters simultaneously in a single tissue scan.

Fornax Technologies (Booth #18)
https://www.fornax-tec.net/en
Fornax Technologies has developed a SPE-workstation based on a 96 well Positive Pressure Extractor (PPE) integrated in a Tecan liquid handler for e.g. MS smart sample preparation. Fornax Technologies utilize refurbished liquid handlers to realize a full automated, reliable, cost-effective SPE-process. Refurbishment, maintenance and service are offered based on certificated processes. The liquid handler is controlled by Graphical user interface portraying the whole SPE-process. Advantages over vacuum manifold: • PPE allows flow pressure regulation according to the application needs • No issues due to leaks of vacuum sealing • No masking of unused wells. Advantages over centrifugation: • No unreliable loading/unloading of centrifuge • Faster processing. The system allows simple automation of sample isolation or extraction, concentration and buffer exchange processes with kits of various suppliers e.g. Waters, Macherey-Nagel, Qiagen, Tecan, GE Healthcare. Fornax Technologies quality policy: second hand and first class!

Fossil Ion Technology (Mini-Table Row)
https://www.fossiliontech.com
SUPER SESI (our core technology) is used by scientists to monitor drug metabolism/toxicity and to identify disease biomarkers in breath in real-time. It is non-invasive, and it detects up to 2200 different metabolites instantly in every single exhalation. SUPER SESI is an ion source that works with High-Resolution Mass Spectrometry to analyze low volatility metabolites in the gas phase.

Hamilton Bonaduz (Booth #11)
http://www.hamilton.ch
Solutions for automated Assays. Fully automated processing of clinical assays requires single-source solutions that are tailor-made for each particular application. Customers benefit from HAMILTON's state-of-the-art technology, the knowledge of a highly qualified team of specialists and solid experience in planning and implementing total solutions for LC-MS sample preparation and a wide range of other applications. Based on the innovative MICROLAB® liquid handling platforms, HAMILTON also offers ready-to-use standard solutions for a wide range of toxicology applications: (1) Protein precipitation, (2) Solid phase extraction, (3) Automated QC and calibration curve preparation, (4) Direct sample injection, (5) Sample dilution, and (6) Reformatting.

Immundiagnostik (Mini-Table Row)
http://www.immundiagnostik.com
Immundiagnostik AG (www.immundiagnostik.com), founded in 1986 by Dr. Franz Paul Armbruster (CEO), is specialized on the development, production, and world-wide distribution of innovative parameters and detection methods for

laboratory diagnostics and medical research. The main focus is the development of immunological tools, of HPLC and molecular biology methods, and of new applications for mass spectrometry (LC-MS/MS). Immundiagnostik concentrates on the development and production of laboratory diagnostics for the identification of disease risks, for differential diagnosis, and for therapeutic drug monitoring. The company holds a particularly strong portfolio in markers of oxidative stress/anti-aging, gastroenterology and nutrition, skeletal system, and cardio-re-no-vascular system. Immundiagnostik owns more than 35 patents in Europe, the US, Japan, Canada, and Australia, is certified according to DIN EN ISO 13485 and fulfills the requirements of the German Medical Device Regulation and the EU IVD Regulations (98/79 EG).

Merck (Booth #4)
http://www.sigmaaldrich.com/clinical
The life science business of Merck combines strong innovative R&D teams with a global network spanning more than 60 countries and 70 manufacturing sites. Innovations include BioSPME for rapid extraction from biological matrices for LC/MS or direct MS, in-line SPE cartridges for effective extraction and phospholipid removal and the latest developments in monolith and fused core HPLC column technologies, all of which can be seen on booth 4. The company provides a product portfolio of 300,000 products, including over 20,000 reference materials, along with Cerilliant Certified Reference Materials, all with easy access to comprehensive data through one of the most advanced web platforms.

PerkinElmer (Booth #27-28)
http://www.perkinelmer.com
PerkinElmer is the global market leader in neonatal screening, currently serving customers in more than 100 countries. We are a total solution provider offering complete systems based on a broad range of high quality, validated products, including newborn screening kits, consumables, instruments and software. PerkinElmer is proud to offer the QSight 220 for Clinical Research System for use in Clinical Research environments that require sensitivity, robustness and reduced downtime. Come visit us at booth 25 to learn more!

Pharmafluidics (Booth #22)
https://www.pharmafluidics.com
Game-changing technology PharmaFluidics introduces the silicon revolution in chromatography. The perfect order of the micro-Chip Pillar Arrays overcomes the physical limits of any packed bed alternative. The revolutionary redesign of the HPLC separation bed reduces peak dispersion to an absolute minimum. We supply micro-Chip Liquid Chromatography cartridges for use in all standard nano-LC equipment Whether you are analyzing trace amounts of compounds in complex mixtures, or looking for subtle modifications in biological molecules: use plug-and-play µPACTM cartridges to boost your biomarker and life sciences research.

RECIPE Chemicals + Instruments (Booth #17)
https://recipe.de/
Starting business in 1982, RECIPE is one of the leading companies in HPLC and LC-MS/MS diagnostics today. For mass spectrometry, RECIPE offers CE/IVD labelled ClinMass® LC-MS/MS Complete Kits. Furthermore, several reagents such as ClinMass® Optimisation Mixes and Internal Standards, ClinCal® Calibrators and ClinChek® Controls are available for a reliable and standardised LC-MS/MS analysis. All products are developed and produced in our state-of-the-art production plant in Munich. RECIPE is recognised worldwide as a reliable partner for clinical laboratories and is certified by the quality management standard EN ISO 13485.

Restek (Booth #9)
http://www.restek.com
A leading innovator of chromatography solutions for both LC and GC, Restek has been developing and manufacturing columns, reference standards, sample preparation materials, accessories, and more since 1985. We provide analysts around the world with products and services to monitor the quality of air, water, soil, food, pharmaceuticals, chemicals, and petroleum products. Our experts have diverse areas of specialization in chemistry, chromatography, engineering, and related fields as well as close relationships with government agencies, international regulators, academia, and instrument manufacturers. www.restek.com

SCIEX (Booth #6)
http://www.sciex.com
SCIEX's global leadership and world-class service and support in the capillary electrophoresis and liquid chromatography-mass spectrometry industry have made it a trusted partner to thousands of the scientists and lab analysts worldwide who are focused on basic research, drug discovery and development, food and environmental testing, forensics and clinical research. As part of AB SCIEX, SCIEX Diagnostics brings the power, flexibility, reliability, and accuracy of mass spectrometry technology to clinical testing laboratories. Offering an expanding portfolio of mass spectrometry based solutions and assays for in vitro diagnostic use, SCIEX Diagnostics enables customers to deliver high quality diagnostic information to clinicians who make decisions affecting patient care.

Shimadzu Booth #1-3
https://www.shimadzu.eu
Since its creation in 1875, Shimadzu has been a worldwide leading manufacturer of analytical instrumentation. Our technologies are constantly driven by innovative thinking and new ideas, developing tools to deliver meaningful data and to help create better science. In the clinical research field, our mass spectrometry platforms in GC-MS, GC-MS/MS, LC-MS/MS and MALDI-TOF do more by bringing together robust, reliable detection together with UFMS (Ultra-Fast Mass Spectrometry) opening up new opportunities in identifying/quantifying more compunds with greater accuracy and certainty. At MSACL 2019 EU, we will show our technologies and workflows for clinical-based research including the new benchtop MALDI-8020. Visit our booth and attend our vendor seminar to discover all the benefits of UFMS technologies!

Tecan (Booth #26)
http://www.tecan.com
Tecan (www.tecan.com) is a leading global provider of laboratory instruments and solutions in biopharmaceuticals, forensics and clinical diagnostics. The company specializes in the development, production and distribution of automated workflow solutions for laboratories in the life sciences sector. Its clients include pharmaceutical and biotechnology companies, university research departments, forensic and diagnostic laboratories. As an original equipment manufacturer (OEM), Tecan is also a leader in developing and manufacturing OEM instruments and components that are then distributed by partner companies. Founded in Switzerland in 1980, the company has manufacturing, research and development sites in both Europe and North America and maintains a sales and service network in 52 countries. In 2017, Tecan generated sales of CHF 548 million (USD 560 million; EUR 494 million). Registered shares of Tecan Group are traded on the SIX Swiss Exchange (TECN; ISIN CH0012100191).

Thermo Fisher Scientific (Booth #14-16)
http://www.thermoscientific.com/msacleu
Thermo Fisher Scientific Inc. is the world leader in serving science, with revenues of $17 billion and approximately 50,000 employees in 50 countries. Our mission is to enable our customers to make the world healthier, cleaner and safer. We help our customers accelerate life sciences research, solve complex analytical challenges, improve patient diagnostics and increase laboratory productivity. Through our premier brands – Thermo Scientific, Applied Biosystems, Invitrogen, Fisher Scientific and Unity Lab Services – we offer an unmatched combination of innovative technologies, purchasing convenience and comprehensive support. For more information, please visit www.thermofisher.com.

UTAK Laboratories (Booth #25)
http://www.utak.com
At UTAK, we're proud to call ourselves "control freaks", but not in the way you might think. That's because our obsession lies not in taking control but in giving control—to the testing labs that need the finest quality control materials for their clinical and forensic toxicology test methods. Our close-knit group crafts the quality controls these labs depend upon for every kind of analysis, including a wide range of comprehensive stock controls in urine, serum, blood, oral fluid and more, as well as starting matrices for laboratories seeking to develop in-house quality control material. We also create personalized control solutions to support the new methods these labs develop. Our dedication is grounded in our belief that better control for testing labs leads to more accurate results and ultimately, to better safeguarding of health and safety standards.

Waters (Booth #12-13)
http://www.waters.com/waters/eventInstance.htm?eiid=134989704
Waters Corporation, the premium brand in the analytical instruments industry, creates business advantages for laboratory-dependent organizations by delivering practical and sustainable scientific innovation to enable significant advancements in healthcare delivery, environmental management, food safety, and water quality worldwide. Bringing keen understanding and deep experience to those responsible for laboratory infrastructure and performance, Waters helps customers make profound discoveries, optimize laboratory operations, deliver product performance, and ensure regulatory compliance. Pioneering a connected portfolio of separations and analytical science, laboratory informatics, mass spectrometry, as well as thermal analysis, Waters' technology breakthroughs and laboratory solutions provide an enduring platform for customer success.

Exhibit Hall Map

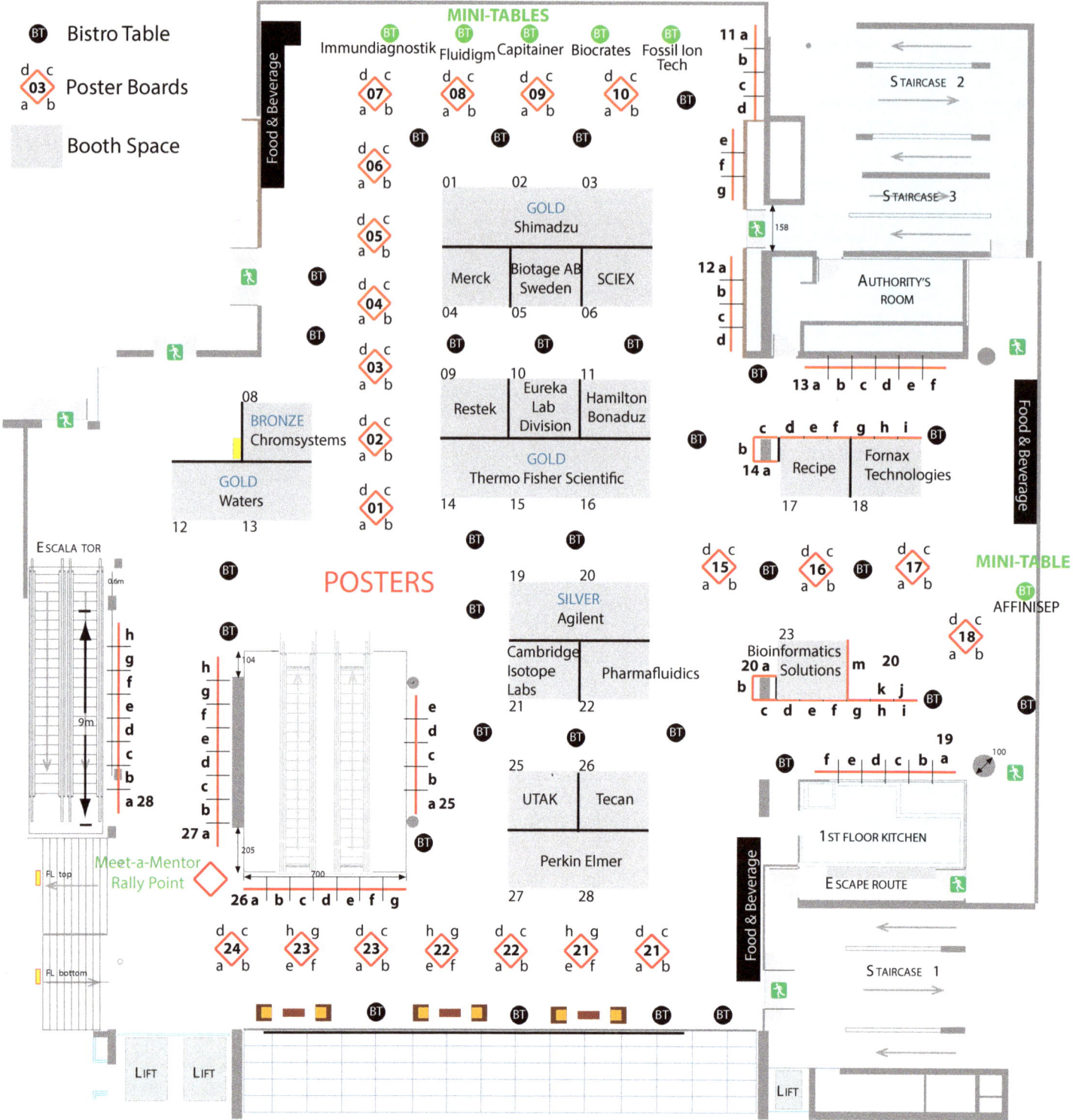

BT Bistro Table

03 Poster Boards
d c / a b

Booth Space

MINI-TABLES

BT Immundiagnostik BT Fluidigm BT Capitainer Biocrates BT Fossil Ion Tech

Food & Beverage

d c **07** a b
d c **08** a b
d c **09** a b
d c **10** a b

BT

11 a
b
c
d

e
f
g

d c **06** a b

BT

BT

BT

158

STAIRCASE 2

STAIRCASE 3

d c **05** a b

| 01 | 02 | 03 |
GOLD
Shimadzu

| Merck | Biotage AB Sweden | SCIEX |
04 05 06

12 a
b
c
d

AUTHORITY'S ROOM

BT

d c **04** a b

BT

BT

BT

BT

13 a b c d e f

BT

d c **03** a b

| 09 | 10 | 11 |
| Restek | Eureka Lab Division | Hamilton Bonaduz |

c d e f g h i
b
14 a Recipe | Fornax Technologies

Food & Beverage

08
BRONZE
Chromsystems

d c **02** a b

GOLD
Thermo Fisher Scientific
14 15 16

17 18

GOLD
Waters
12   13

d c **01** a b

14 15 16

d c **15** a b
BT
d c **16** a b
BT
d c **17** a b

MINI-TABLE

BT

BT AFFINISEP

POSTERS

ESCALATOR
0.6m

BT

BT

19   20
SILVER
Agilent

d c **18** a b

9m

h
g
f
e
d
c
b
a **28**

104

h
g
f
e
d
c
b
a

27 a

205

e
d
c
b
a **25**

BT

Cambridge Isotope Labs
21

Pharmafluidics
22

BT

BT

BT

23
Bioinformatics Solutions
20 a
b

m **20**
k j

c d e f g h i

BT

700

BT

19
f e d c b a

100

Meet-a-Mentor Rally Point

FL top

FL bottom

26 a b c d e f g

25 26
UTAK | Tecan

Perkin Elmer

27   28

1ST FLOOR KITCHEN

ESCAPE ROUTE

d c **24** a b
h g **23** e f
d c **23** a b
h g **22** e f
d c **22** a b
h g **21** e f
d c **21** a b

Food & Beverage

STAIRCASE 1

BT BT BT BT

LIFT LIFT

LIFT

Session Room Map

5th Floor

Out elevator and straight back to end of hall

Trapp
5th Floor

4th Floor

Track 6
Doppler
4th Floor

3rd Floor

Track 5
Trakl
3rd Floor

2nd Floor

Track 4
Paracelsus
2nd Floor

Ground Floor :: Tracks 1-3

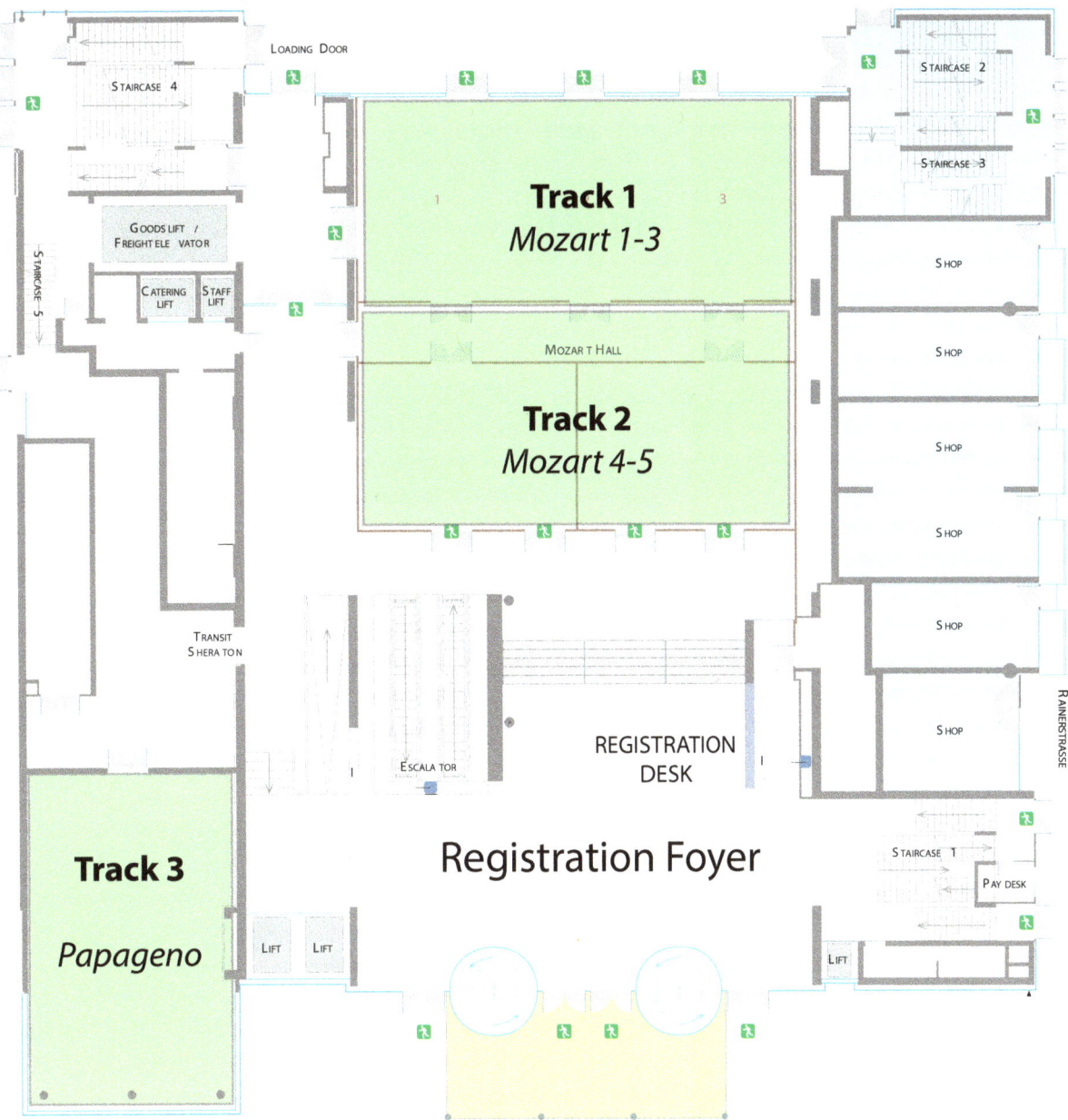

LOADING DOOR

STAIRCASE 4

STAIRCASE 2

STAIRCASE 3

GOODS LIFT / FREIGHT ELEVATOR

STAIRCASE 5

CATERING LIFT

STAFF LIFT

Track 1
Mozart 1-3

1   3

MOZART HALL

Track 2
Mozart 4-5

SHOP

SHOP

SHOP

SHOP

SHOP

SHOP

TRANSIT SHERATON

ESCALATOR

REGISTRATION DESK

RAINERSTRASSE

Registration Foyer

STAIRCASE 1

PAY DESK

Track 3
Papageno

LIFT   LIFT

LIFT

433   433   433

Main Entrance

Exhibitors

GOLD SPONSOR LEVEL

Thermo Fisher SCIENTIFIC

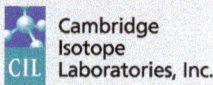

SHIMADZU Excellence in Science

Waters THE SCIENCE OF WHAT'S POSSIBLE.™

SILVER SPONSOR LEVEL

Agilent

BRONZE SPONSOR LEVEL

CHROMSYSTEMS DIAGNOSTICS BY HPLC & LC-MS/MS

And special thanks to

Thermo Fisher SCIENTIFIC

CIL Cambridge Isotope Laboratories, Inc.

for their continued support of the Educational Grant Program

AFFINISEP
Biocrates Life Sciences
Bioinformatics Solutions
Biotage AB, Sweden
Cambridge Isotope Labs
Capitainer
Eureka Lab Division
Fluidigm
Fornax Technologies
Fossil Ion Technology

Hamilton Bonaduz
Immundiagnostik
Merck
PerkinElmer
Pharmafluidics
RECIPE Chemicals + Instr
Restek
SCIEX
Tecan
UTAK Laboratories

Supporting Societies

AACC

EFLM EUROPEAN FEDERATION OF CLINICAL CHEMISTRY AND LABORATORY MEDICINE

RNE Portuguese Mass Spectrometry Network

Belgian Society for Mass Spectrometry

SPB Sociedade Portuguesa de Bioquímica

CZECH MS Czech Society for Mass Spectrometry

ФЕДЕРАЦИЯ ЛАБОРАТОРНОЙ МЕДИЦИНЫ

Clinical Mass Spectrometry
MSACL 2020 US
12th Annual Conference & Exhibits

Palm Springs, CA Mar 31- Apr 2
Short Courses March 29-31

Call for Abstracts : Podium by Nov 13

Indian Canyons, Palm Springs

Education • Networking • Mentoring

Call for Abstracts for Scientific Sessions on:

Also accepting **Case Studies** on all topics.

Microbiology
Applications of MALDI MS
New & Experimental Approaches
Anti-Microbial Resistance

Tox / TDM / Endo
Clin Apps of TDM & Pharmacokinetics
Cannabis / Hemp / Natural Products
Urine & Oral Fluid Drug Testing
Forensic Tox & Drug ID
Functional Assays & HTS

Metabolomics
Translation
Lipidomics
Large Cohorts
Multi-omic Analysis
Untargeted Profiling
New Techniques & Approaches

Data Science
Machine Learning
Practical Tools
R Showcase
Best Practices

Proteomics
Data-independant Acquisition
Reference Method Procedures
Post-trans Modifications
Beyond Serum/Plasma
Therapeutic Antibodies
Tissue (Quantitative)
Microsampling
Intact Proteins
New Assays

Imaging MS
Biomarkers
In the Clinic
Informatics
Translational Research
Therapeutic Targets
Applications & Strategies
Imaging Hardware / Tech

MSACL
Est. 2008
Clinical Mass Spectrometry
m/z

Podium due by Nov 13
Poster due by Feb 20

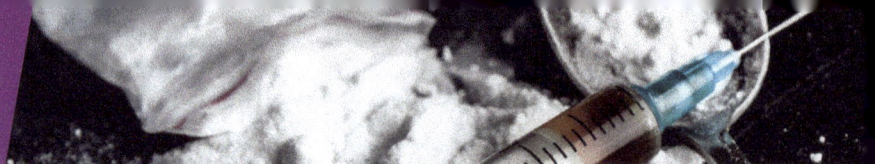

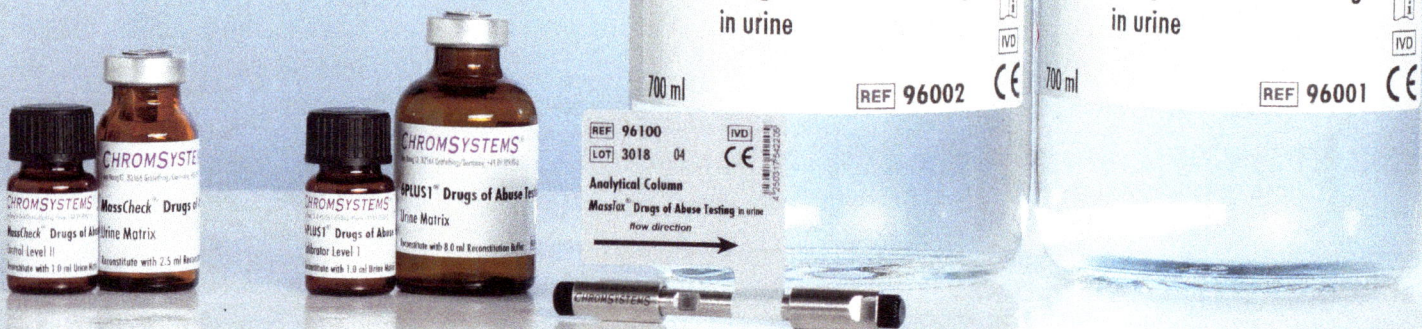

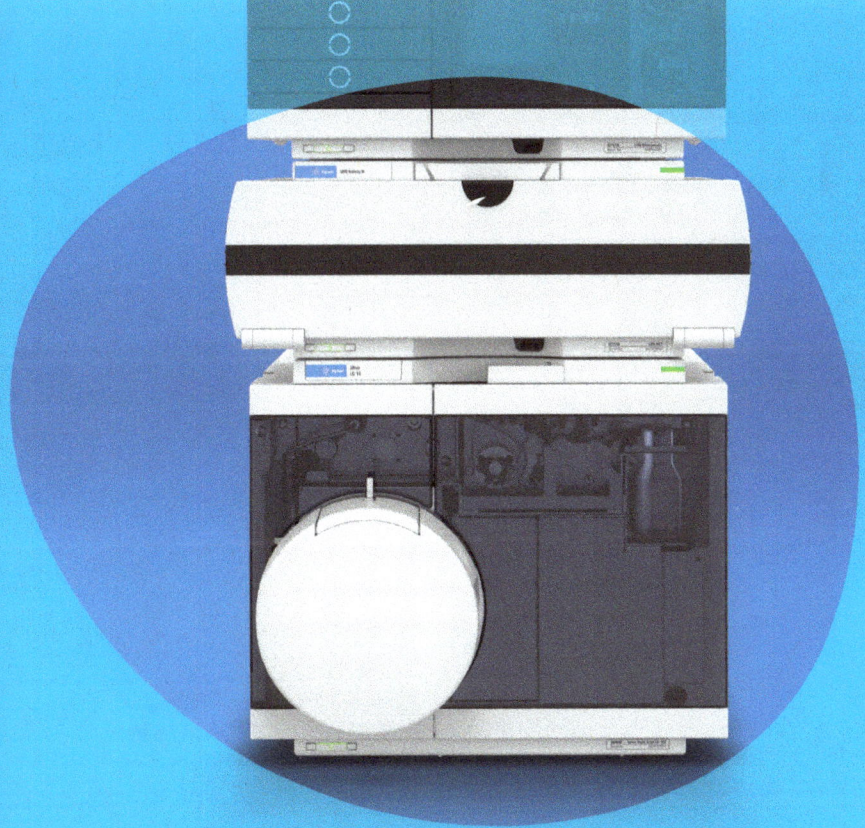

Unbelievably Powerful, Remarkably Small

Take your clinical research to new heights with Agilent's affordable, space-saving Ultivo LC/MS Triple Quad.

For starters, the Ultivo offers more flexibility with the new electrospray ionization (ESI) configuration.

You can also take advantage of enhanced data integrity tools—plus features that help your lab comply with 21 CFR, Part 11 requirements. And with new early maintenance feedback capabilities, Ultivo is even more user friendly.

With more options than ever—Ultivo is the clear choice for your lab.

See how Ultivo enables precise, high-throughput quantitation: www.agilent.com/chem/Ultivo

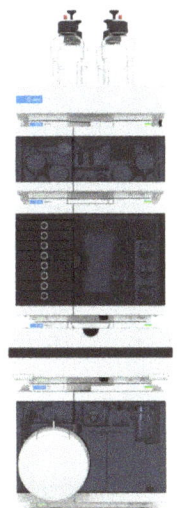

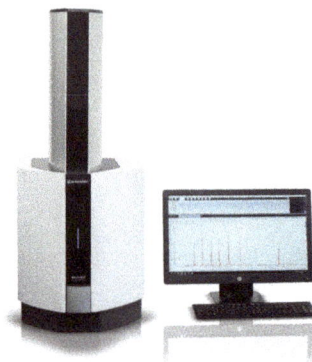